CW01510638

Oxford Lecture Series in
Mathematics and its Applications 2

*Series editors*
John Ball   Dominic Welsh

## OXFORD LECTURE SERIES IN
## MATHEMATICS AND ITS APPLICATIONS

1. John C. Baez (ed): *Knots and quantum gravity*
2. Irene Fonseca and Wilfrid Gangbo: *Degree theory in analysis and applications*

# Degree Theory in Analysis and Applications

Irene Fonseca

and

Wilfrid Gangbo

*Department of Mathematics*
*Carnegie Mellon University*

CLARENDON PRESS · OXFORD
1995

*Oxford University Press, Walton Street, Oxford OX2 6DP*

*Oxford New York*
*Athens Auckland Bangkok Bombay*
*Calcutta Cape Town Dar es Salaam Delhi*
*Florence Hong Kong Istanbul Karachi*
*Kuala Lumpur Madras Madrid Melbourne*
*Mexico City Nairobi Paris Singapore*
*Taipei Tokyo Toronto*
*and associated companies in*
*Berlin Ibadan*

*Oxford is a trade mark of Oxford University Press*

*Published in the United States by*
*Oxford University Press Inc., New York*

© *I. M. Q. C. da Fonseca and W. Gangbo, 1995*

*All rights reserved. No part of this publication may be*
*reproduced, stored in a retrieval system, or transmitted, in any*
*form or by any means, without the prior permission in writing of Oxford*
*University Press. Within the UK, exceptions are allowed in respect of any*
*fair dealing for the purpose of research or private study, or criticism or*
*review, as permitted under the Copyright, Designs and Patents Act, 1988, or*
*in the case of reprographic reproduction in accordance with the terms of*
*licences issued by the Copyright Licensing Agency. Enquiries concerning*
*reproduction outside those terms and in other countries should be sent to*
*the Rights Department, Oxford University Press, at the address above.*

*This book is sold subject to the condition that it shall not,*
*by way of trade or otherwise, be lent, re-sold, hired out, or otherwise*
*circulated without the publisher's prior consent in any form of binding*
*or cover other than that in which it is published and without a similar*
*condition including this condition being imposed*
*on the subsequent purchaser.*

*A catalogue record for this book is available from the British Library*

*Library of Congress Cataloging in Publication Data*
*(Data applied for)*

*ISBN 0 19 851196 5*

*Typeset by the authors using LaTeX*

*Printed in Great Britain by*
*Bookcraft (Bath) Ltd*
*Midsomer Norton, Avon*

# PREFACE

In these lecture notes we study degree theory and some of its applications in analysis. We will focus on the recent developments of this theory for Sobolev functions which cannot be found in the existing literature dealing with the notion of topological degree for continuous functions (see e.g. Deimling 1985, Gold'sthein and Reshetnyak 1990, Lloyd 1978, and Schwartz 1969). In recent years, the need to extend the notion of degree to nonsmooth functions was motivated in part by applications to nonlinear analysis and, specifically, to nonlinear elasticity (see Ball 1978, Fonseca and Gangbo 1995, Müller *et al.* 1994, Šverák 1988, and Tang 1988). As an example, in Chapter 6 we illustrate how a change of variables formula for Sobolev functions yields a local invertibility theorem for Sobolev functions which, in turn, allows us to prove a lower semicontinuity theorem for energy functionals involving variation of the domain.

Our primary goal was to assemble the literature that is nowadays scattered on papers, manuscripts, and sometimes private communications, hopefully rendering these results readily accessible to analysts with a reasonable background in infinitesimal real analysis.

In order to keep these notes as self contained as possible, in Chapter 1 and following the work of Schwartz (1969) we define the degree for differentiable functions and then extend the definition to continuous functions.

In Chapter 2 we give some properties of the degree for continuous functions and we study the continuity of $d(\phi, D, p)$ with respect to $\phi, D, p$. We show that the degree $d(\phi, D, p)$ gives a topological characterization of $\phi|_{\partial D}$. Indeed, let $B^N = \{x \in \mathbb{R}^N : |x| \leq 1\}$ and let $S^{N-1}$ be the boundary of $B^N$. If $\varphi : S^{N-1} \to S^{N-1}$ is a continuous mapping, then the following assertions are equivalent (see Corollary 2.13):

$\varphi$ is not homotopic to a constant.

Every continuous extension $\phi : \bar{B}^N \to \mathbb{R}^N$ admits a zero.

Every continuous extension $\phi : \bar{B}^N \to \mathbb{R}^N$ verifies $d(\phi, B^N, 0) \neq 0$.

In Chapter 3 we present some classical results as applications of degree theory in topology, namely the Brouwer Fixed Point Theorem, the Borsuh–Ulam Theorem, the Jordan Separation Theorem, the Invariance of Domain Theorem (Open Mapping Theorem), the Perron–Frobenuis Theorem.

In Chapter 4 we briefly review the theory of Sobolev spaces, introducing the necessary background for Chapter 5, where we give some properties of degree for Sobolev functions $\phi \in W^{1,p}$. Following Marcus and Mizel (1973), Gold'sthein and Reshetnyak (1990), and Šverák (1988), we state and prove change of variables formulae using the notion of degree. The case $p > N$ is due to Marcus and Mizel (1973), the case $p = N$ was treated by Gol'dshtein and Reshetnyak (1990), and the case $N - 1 < p < N$ was first studied by Šverák (1988). These change of variables formulae rely on the fact that $d(\phi, B(x, r), \phi(x)) = \text{sgn}(\det \nabla\phi(x))$ for $r$ small enough and whenever $\det \nabla\phi(x) \neq 0$.

In Chapter 6 we give some applications of degree theory in Sobolev spaces and we prove a local inverse function theorem for $u \in W^{1,N}(D)^N$ under the condition $\det \nabla u(x) > 0$ for almost every $x \in D \subset \mathbb{R}^N$ (see Fonseca and Gangbo 1995).

Finally, in Chapter 7 we extend the notion of degree to infinite-dimensional spaces. We prove that most of the properties of degree discussed in Chapter 3 still hold in this context and we present some applications of degree theory to ordinary and partial differential equations.

This work was supported by the Army Research Office and the National Science Foundation through the Center for Nonlinear Analysis at Carnegie Mellon University. Also, the work of the first author, Irene Fonseca, was partially supported by the National Science Foundation under Grant No. DMS-9201215.

The authors would like to thank J. Ball and J. Manfredi for their very hepful suggestions. Our thanks also go to Mrs E. Gangbo who skilfully typed the manuscript.

Pittsburgh                                                                        I.F., W.G.
March 1995

# CONTENTS

**Introduction**     1

**1    Degree theory for continuous functions**     5
1.1   Topological degree for $C^1$ functions     5
1.2   Topological degree for continuous functions     16
1.3   Generalization of the degree     20
1.4   Exercises     25

**2    Degree theory in finite-dimensional spaces**     30
2.1   Dependence of the degree on $\phi$ and $p$     30
2.2   Dependence of the degree on the domain $D$     32
2.3   The multiplication theorem     35
2.4   An application of Hopf's theorem     39
2.5   Degree and winding number     41
2.6   Exercises     46

**3    Some applications of the degree theory to topology**     48
3.1   The Brouwer Fixed Point Theorem     48
3.2   Odd mappings     54
3.3   The Jordan Separation Theorem     64
3.4   Exercises     71

**4    Measure theory and Sobolev spaces**     74
4.1   Review of measure theory     74
4.2   Hausdorff measures     78
4.3   Overview of Sobolev spaces     87
4.4   $p$-capacity     92

**5    Properties of the degree for Sobolev functions**     106
5.1   Results of weakly differential mappings     107
5.2   Weakly monotone functions     119
5.3   Change of variables via the multiplicity function     131
5.4   Change of variables via the degree     135
5.5   Change of variables for Sobolev functions     140

**6    Local invertibility of Sobolev functions and applications**     149
6.1   Local invertibility in $W^{1,N}$     149
6.2   Energy functionals involving variation of the domain     160

**7 Degree in infinite-dimensional spaces** 172
7.1 Introduction to the Leray–Schauder degree 172
7.2 Properties of the Leray–Schauder degree 177
7.3 Fixed point theorems 185
7.4 An application of the degree theory to ODEs 190
7.5 First application of the degree theory to PDEs 192
7.6 Second application of the degree theory to PDEs 199
7.7 Exercises 203

**References** 205

**Index** 209

# INTRODUCTION

The degree $d(\phi, D, p)$ is a tool that describes the number of solutions for the equation $\phi(x) = p$ in a given open set $D \subset X$, where $\phi : D \subset X \to X$ is a continuous function, $p \notin \phi(\partial D)$, and $X$ is a topological space, often a metric space. As it turns out, the degree is a generalization of the winding number of $\phi : \bar{D} \subset \mathbb{C} \to \mathbb{C}$ when we identify the complex plane $\mathbb{C}$ to $\mathbb{R}^2$ (see Chapter 2). Assuming that

$$A \subset \{(\phi, D, p) \; : \; \phi : D \to X \text{ is continuous, } p \notin \phi(\partial D)\},$$

we search for a function $d : A \to \mathbb{Z}$ satisfying the following properties:

(D1) If $X = \mathbb{R}^N$, $D$ is open, bounded and if $p \in D$, then

$$d(I|_D, D, p) = 1,$$

where $I$ denotes the identity mapping of $X$.

(D2) If $d(\phi, D, p) \neq 0$, then there exists $x \in D$ such that $\phi(x) = p$.

(D3) If $D_1 \cap D_2 = \emptyset$ and if $p \notin \phi(\partial D_1 \cup \partial D_2)$, then

$$d(\phi|_{D_1}, D_1, p) + d(\phi|_{D_2}, D_2, p) = d(\phi, D_1 \cup D_2, p).$$

(D4) If $h : [0, 1] \to C^0(D)^N$ is a $C^0$ homotopy such that $p \notin h(t)(\partial D)$ for all $t \in [0, 1]$, then

$$d(h(t), D, p) = d(h(0), D, p).$$

(D5) If $p \notin \phi(\partial D)$, then

$$d(\phi, D, p) = d(\phi - p, D, 0).$$

There is only one function $d : A \to \mathbb{Z}$ verifying (D1)–(D5) (see Lloyd 1978 and Deimling 1985) and this is called the *topological degree*. In finite dimensions it is also designated by the *Brouwer degree* since the idea was first developed by Brouwer in 1912.

There are many approaches to the introduction of the notion of degree and the background involved in each one of them may differ considerably. As an example, Alexandroff and Hopf (1935) and Dold (1972) made considerable use of concepts from algebraic topology and group theory, while the approach that we adopt here is more recent and uses only some basic analytical tools, such as the Implicit Function Theorem and Sard's Lemma. The analytic point of view was first given by Nagumo (1951).

There is a very strong connection between the notions of topological degree and null Lagrangians (see Ball 1978). This was remarked upon by Tartar in 1974

1

(see Tartar 1988). Null Lagrangians, such as $\det \nabla \phi$ and lower order minors of $\nabla \phi$, depend only on the trace values of $\phi$ on $\partial D$ and are thus weakly continuous since the oscillations that may take place inside $D$ are not felt by the functional

$$\phi \to \int_D M(\nabla \phi(x)) \, dx,$$

where $M(\nabla \phi)$ denotes the list of all minors of $\nabla \phi$.

As mentioned earlier, the degree $d(\phi, D, p)$ is sought to count the number of solutions for the equation $\phi(x) = p, p \notin \phi(\partial D)$. Let us assume that $D$ is an open, bounded, smooth subdomain of $\mathbb{R}^N$. In the scalar case, we consider $D = (0, L), \phi : (0, L) \to \mathbb{R}, p \notin \{\phi(0), \phi(L)\}$. Note that information on the boundary values of $\phi$ may help, in same cases, to decide whether or not $\phi(x) = p$ admits a solution. Indeed, if

$$(\phi(0) - p) \cdot (\phi(L) - p) < 0,$$

then there is at least one zero of $\phi(x) - p$ in $D$. We may count the number of such solutions as

$$d(\phi, D, p) = \begin{cases} 1 & \text{if } \phi(0) < p < \phi(L) \\ -1 & \text{if } \phi(0) > p > \phi(L) \\ 0 & \text{otherwise.} \end{cases}$$

In turn, this formula reduces to

$$d(\phi, D, p) = \sum_{x \in \phi^{-1}(p)} \operatorname{sgn}(\phi'(x)), \tag{0.1}$$

whenever $\phi \in C^1$ and $\phi^{-1}(\{p\}) \subset \{x \in D : \phi'(x) \neq 0\}$. This is in agreement with Definition 1.2. In order to extend this notion to higher dimensions, let $\phi \in C^1(\bar{D})^N, v \in C(\mathbb{R}^N)$ and define

$$I_v(\phi) := \int_D v(\phi(x)) J_\phi(x) \, dx,$$

where $J_\phi(x) := \det \nabla \phi(x)$. We show that, for $\psi \in C^1(\bar{D})^N$,

$$I_v(\phi) = I_v(\psi) \qquad \text{if } \phi = \psi \text{ on } \partial D. \tag{0.2}$$

Heuristically, this can easily be seen in the case where $\phi$ and $\psi$ are invertible, as, then, using the change of variables formula, we would have

$$I_v(\phi) = \int_{\phi(D)} v(y) \, dy \tag{0.3}$$

$$= \int_{\psi(D)} v(y) \, dy$$

$$= I_v(\psi).$$

In general, we assume first that $v \in C^1(\mathbb{R}^N)$, $\phi, \psi \in C^2(\bar{D})^N$ and we remark that

$$\frac{d}{dt} I_v(\phi + t(\psi - \phi)) = \int_{\partial D} v(\phi + t(\psi - \phi)) \, \xi_{H_0(x,t)} \cdot \nu \, dH^{N-1}(x), \qquad (0.4)$$

where $H^{N-1}$ stands for the $N-1$-dimensional Hausdorff measure, $H_0(x,t) := \phi(x) + t(\psi(x) - \phi(x))$ and the $k$th component of $\xi_{H(x,t)}$, for a given function $H(x,t)$, is defined by

$$\left(\xi_{H(x,t)}\right)_k :=$$
$$\det\left(\frac{\partial H}{\partial x_1}(x,t), \ldots, \frac{\partial H}{\partial x_{k-1}}(x,t), \frac{\partial H}{\partial t}(x,t), \frac{\partial H}{\partial x_{k+1}}(x,t), \ldots, \frac{\partial H}{\partial x_N}(x,t)\right).$$

Thus, if $\psi = \phi$ on $\partial D$ we obtain (0.2) for smooth $v, \phi, \psi$ and the result now follows for $v \in C(\mathbb{R}^N)$, $\phi, \psi \in C^1(\bar{D})^N$ via a density argument.

Using (0.4) it can also be seen that if $\phi, \psi$ are homotopic (see Definition 1.11), precisely if there exists a smooth function $H : \bar{D} \times [0,1] \to \mathbb{R}^N$ such that

$$H(x,0) = \phi(x), \ H(x,1) = \psi(x) \qquad (0.5)$$

and

$$v(H(x,t)) = 0 \text{ whenever } (x,t) \in \partial D \times [0,1],$$

then

$$I_v(\phi) = I_v(\psi). \qquad (0.6)$$

It suffices to perform the differentiation

$$\frac{d}{dt} I_v(H(\cdot,t)) = \int_{\partial D} v(H(x,t)) \, \xi_{H(x,t)} \cdot \nu \, dH^{N-1}(x) = 0.$$

Assume that $\phi \in C(\bar{D})^N$ and that $\phi(\partial D) \cap \operatorname{spt} v = \emptyset$. It is possible to find $\phi_n \in C^1(\bar{D})^N$ such that

$$\phi_n \to \phi \text{ uniformly on } \bar{D}$$

and

$$\phi_n, \phi_m \text{ are homotopic in the sense of (0.5)}$$

for $n, m$ large enough. Hence by (0.6) we have

$$I_v(\phi_n) \text{ constant for } n \text{ large}$$

and we set

$$I_v(\phi) := \lim_{n \to +\infty} I_v(\phi_n). \tag{0.7}$$

Based on this property, if $\phi \in C(\bar{D})^N$ and if $p \notin \phi(\partial D)$, then for $n$ large enough,

$$\phi(\partial D) \cap \operatorname{spt} v_n = \emptyset,$$

where

$$v_n(y) := n^N \rho(n(y-p)), \rho \in C_c^1(\mathbb{R}^N), \int_{\mathbb{R}^N} \rho(y)\,dy = 1, \operatorname{spt}\rho = \bar{B}(0,1),$$

and so, due to (0.7), $I_{v_n}(\phi)$ is well defined and we set

$$d(\phi, D, p) := \lim_{n \to +\infty} I_{v_n}(\phi). \tag{0.8}$$

We remark that this is in agreement with Proposition 1.7. From (0.2), (0.7), and (0.8), it follows that

$$d(\phi, D, p) = d(\psi, D, p) \text{ if } \phi = \psi \text{ on } \partial D$$

(see also Theorem 2.4).

Using Sard's Lemma 1.4, it is also possible to show that if $\phi \in C^1(\bar{D})^N$ and if $p \notin \phi(\partial D)$, $\phi^{-1}(\{p\}) \subset \{x \in D : J_\phi(x) \neq 0\}$, then (0.1) becomes

$$d(\phi, D, p) = \sum_{x \in \phi^{-1}(p)} \operatorname{sgn}(J_\phi(x)). \tag{0.9}$$

Properties (D1)–(D5) follow from the definition of topological degree (see also Theorems 2.1, 2.3, Proposition 2.5, and Theorem 2.7). In addition, degree is a constant integer on every connected component of $D \setminus \phi(\partial D)$ (see Theorem 2.3), and the change of variables formula (0.3) can now be written as (see Theorem 5.31)

$$I_v(\phi) = \int_D v(\phi(x)) \det \nabla \phi(x)\,dx$$
$$= \int_{\mathbb{R}^N} v(y) d(\phi, D, y)\,dy.$$

Often, to determine the degree $d(\phi, D, p)$ we construct a homotopy between $\phi$ and a simple $C^1$ function to which formula (0.9) applies and we use the invariance property of the degree under homotopies. Such is the case of Lemma 5.9, where to show that a continuous function, differentiable at $x_0$, with $J_\phi(x_0) \neq 0$, has degree equal to $\operatorname{sgn}(J_\phi(x_0))$ near $x_0$, we construct a homotopy between $\phi$ and the linear function

$$\psi(x) = \phi(x_0) + \nabla\phi(x_0)(x - x_0).$$

This result is used later in Chapter 6, Lemma 6.5.

# 1

# DEGREE THEORY FOR CONTINUOUS FUNCTIONS

In the first three chapters, unless the contrary is explicitly stated, $D$ is a bounded, open subset of $\mathbb{R}^N$. Also, in the sequel we will use the following notation:

- $x = (x_1, \ldots, x_N) \in \mathbb{R}^N$, the infinity norm of $x$ is $|x| := \max\{|x_i| : i = 1, \ldots, N\}$, and the euclidean norm of $x$ is $|x|_2 := (x_1^2, \ldots, x_N^2)^{\frac{1}{2}}$. We set $\rho(x, y) := |x - y|$ and $\operatorname{dist}(x, y) := |x - y|_2$. If $x = (x_1, \ldots, x_N) \in \mathbb{R}^N$ and $y = (y_1, \ldots, y_N) \in \mathbb{R}^N$, we denote by $x \cdot y$ the inner product $\sum_{i=1}^N x_i y_i$.
- We identify the set $\mathbb{R}^{M \times N}$ of all $M \times N$ matrices with $\mathbb{R}^{MN}$ and if $A$ is an $M \times N$ matrix we denote by $|A|$ (resp. $|A|_2$) the infinity norm (resp. the euclidean norm) of $A$.
- If $S \subset \mathbb{R}^N$, the *distance of $x$ to $S$* is given by $\inf\{\rho(x, y) : y \in S\}$ and is denoted by $\rho(x, S)$.
- $Q(x, r) := \{y \in \mathbb{R}^N : \rho(x, y) < r\}$.
- $B(x, r) := \{y \in \mathbb{R}^N : \operatorname{dist}(x, y) < r\}$.
- If $S \subset \mathbb{R}^N, \bar{S}$ stands for the closure of $S$, int $S$ its interior, $\partial S$ its boundary, and $S^c$ its complement.
- $C(\bar{D})^N := \{f : \bar{D} \to \mathbb{R}^N : f \text{ is continuous}\}$ and $\|f\| := \sup\{|f(x)| : x \in D\}$.
- $C_c(D)^N := \{f \in C(\bar{D})^N : \operatorname{spt}(f) \subset\subset D\}$, where we write $A \subset\subset B$ if $\bar{A}$ is a compact subset of $B$.
- $C_0(\mathbb{R}^N)^N := \{f \in C(\mathbb{R}^N)^N : \lim_{|x| \to \infty} |f(x)| = 0\}$.
- If $f \in C^1(D)^N, \nabla f(x) = \left(\frac{\partial f_i}{\partial x_j}(x)\right)_{i,j=1,\ldots,N}$, and $J_f(x) := \det \nabla f(x)$.
- $C^1(\bar{D})^N$ denotes the set of functions $f \in C(\bar{D})^N$ such that $f$ admits an extension $\tilde{f}$ to an open set $D(f)$ containing $\bar{D}$, and $\nabla \tilde{f}$ is continuous on $D(f)$.
- If $f \in C^1(\bar{D})^N$, then $\|f\|_1 := \|f\| + \|\nabla f\|$.
- If $p \in \mathbb{R}^N, x \in \bar{D}$ is said to be a *$p$-point of $\phi$* if $\phi(x) = p$.
- $\mathcal{L}^N$ stands for the Lebesgue measure in $\mathbb{R}^N$ (see Definition 4.1).
- $n(A)$ denotes the number of distinct elements of $A$ when $A$ is a finite set, and

$$\sharp(A) := \begin{cases} n(A) & \text{if } A \text{ is a set of finite cardinality} \\ +\infty & \text{if } A \text{ is a set of infinite cardinality.} \end{cases}$$

## 1.1 Topological degree for $C^1$ functions

**Definition 1.1** *Let $\phi \in C^1(\bar{D})^N, x \in \bar{D}$. We say that $x$ is a critical point of $\phi$ if $J_\phi(x) = 0$. In this case $\phi(x)$ is a critical value of $\phi$. We define $Z_\phi := \{x \in$*

$\bar{D} : J_\phi(x) = 0\}$, and $\phi(Z_\phi)$ is called the crease of $\phi$. If $p \notin \phi(Z_\phi)$, then $p$ is said to be a regular value of $\phi$.

**Definition 1.2** Let $\phi \in C^1(\bar{D})^N$ and $p \notin (\phi(Z_\phi) \cup \phi(\partial D))$. The degree of $\phi$ at $p$ with respect to $D$ is defined by

$$d(\phi, D, p) := \sum_{x \in \phi^{-1}(p)} \mathrm{sgn}(J_\phi(x)), \qquad (1.1)$$

where $\mathrm{sgn}(t) = 1$ for $t > 0$ and $\mathrm{sgn}(t) = -1$ for $t < 0$.

Notice that since $p \notin \phi(Z_\phi)$, $\phi^{-1}(p)$ is finite and so (1.1) is well defined (see Exercise 1.2).

The task ahead will be to relax the conditions $p \notin \phi(Z_\phi)$ and $\phi \in C^1(\bar{D})^N$ imposed in Definition 1.2.

**Proposition 1.3** Let $\phi \in C^1(\bar{D})^N$ and $p \notin \phi(Z_\phi) \cup \phi(\partial D)$. Then there exists $\delta > 0$ such that if $\psi \in C^1(\bar{D})^N$ and $||\phi - \psi||_1 \leq \delta$, then $p \notin \psi(Z_\psi) \cup \psi(\partial D)$ and

$$d(\phi, D, p) = d(\psi, D, p).$$

**Proof**  As $\phi^{-1}(p)$ is either finite or empty (see Exercise 1.2), we consider these two cases separately.

Firstly we assume that $\phi^{-1}(p) = \emptyset$. Then $\delta = \frac{1}{2}\rho(p, \phi(\bar{D})) > 0$ and, given $\psi \in C^1(\bar{D})^N$ such that $||\phi - \psi||_1 < \delta$, we have $\psi^{-1}(p) = \emptyset$. Thus $p \notin \psi(Z_\psi) \cup \psi(\partial D)$ and

$$d(\phi, D, p) = d(\psi, D, p) = 0.$$

Next, suppose that $\phi^{-1}(p) = \{a_1, \dots, a_k\}$. We show that there exist $r > 0, \delta > 0$, such that whenever $\psi \in C^1(\bar{D})^N$, $||\phi - \psi||_1 \leq \delta$, then $\psi$ admits exactly one $p$-point in $Q(a_i, r), i = 1, \dots, k$.

Indeed, fix $r_0$ such that

$$0 < r_0 < \min\left\{ \frac{\rho(a_i, a_j)}{3} : i \neq i, \ i, j = 1, \dots, k \right\},$$

$$r_0 < \min\left\{ \frac{\rho(a_i, \partial D \cup Z_\phi)}{3} : i = 1, \dots, k \right\}.$$

Set

$$Q(r) := Q(a_1, r) \cup \dots \cup Q(a_k, r)$$

and

$$c := \min\{|J_\phi(a_i)| : i = 1, \dots, k\}.$$

We have $c > 0$ and since $J_\phi$ is continuous on $\bar{D}$, there exists $0 < r_1 < r_0$ such that $|J_\phi(x)| \geq \frac{2}{3}c$ for every $x \in Q(r_1)$. Choosing $\delta_1 > 0$ such that

$$\sup\{|J_\phi(x) - J_\psi(x)| : x \in \bar{D}\} \leq \frac{1}{3}c$$

whenever $||\phi - \psi||_1 \le \delta_1$, we deduce that

$$\sup\{|J_\psi(x)| : x \in Q(r_1)\} \ge \frac{1}{3}c$$

if $||\phi - \psi||_1 \le \delta_1$. Now we fix $i \in 1, \ldots, k$ and we want to solve the equation $\psi(x) = p$ in $Q(a_i, r_1)$. For simplicity of notation we write

$$a := a_i, \ h := \phi(a_i) - \psi(a_i), V := (\nabla\psi(a))^{-1}.$$

Define $T, W : Q(0, r_1) \to \mathbb{R}^N$ as

$$Tz := \psi(a + z) - \psi(a) - \nabla\psi(a)z, \ W(z) := V(h - Tz).$$

We have

$$\psi(a + z) = p, \quad z \in Q(0, r_1) \Leftrightarrow \psi(a + z) = \phi(a), \quad z \in Q(0, r_1)$$
$$\Leftrightarrow W(z) = z, \quad z \in Q(0, r_1).$$

*Claim 1.* The equation $W(z) = z$ admits exactly one solution in $Q(0, r)$ for some $r < r_1$ and when $||\phi - \psi||_1 < \delta$ for some $\delta < \delta_1$.

In order to show that $W$ is a contraction, we estimate the component $(Tz - Ty)_l$ for $y, z \in Q(0, r)$.

$$(Tz - Ty)_l = \psi_l(a + z) - \psi_l(a + y) - (\nabla\psi(a)(z - y))_l$$
$$= \int_0^1 \frac{d}{d\theta}\psi_l(a + \theta z + (1 - \theta)y)\, d\theta - (\nabla\psi(a)(z - y))_l$$
$$= \int_0^1 \sum_{j=1}^N (z_j - y_j)\left[\frac{\partial\psi_l}{\partial x_j}(a + \theta z + (1 - \theta)y) - \frac{\partial\psi_l}{\partial x_j}(a)\right] d\theta$$
$$= \sum_{j=1}^N (z_j - y_j)\int_0^1 \left[\frac{\partial\psi_l}{\partial x_j}(\zeta) - \frac{\partial\phi_l}{\partial x_j}(\zeta) + \frac{\partial\phi_l}{\partial x_j}(\zeta)\right.$$
$$\left. - \frac{\partial\phi_l}{\partial x_j}(a) + \frac{\partial\phi_l}{\partial x_j}(a) - \frac{\partial\psi_l}{\partial x_j}(a)\right] d\theta,$$

where $\zeta = a + \theta z + (1 - \theta)y$. This implies that

$$|T(z) - T(y)| \le N|z - y|\int_0^1 [2\delta + \epsilon(r)]\, d\theta,$$

where $\epsilon : [0, r_1] \to \mathbb{R}$ is defined as

$$\epsilon(r) := \sup \left\{ \int_0^1 \left| \frac{\partial \phi_i}{\partial x_j}(a + \theta z + (1 - \theta)y) - \frac{\partial \phi_i}{\partial x_j}(a) \right| d\theta : \right.$$
$$\left. y, z \in \bar{Q}(0, r), i, j = 1, \dots, N \right\}.$$

Clearly, $\epsilon$ is a nondecreasing function and $\lim_{r \to 0+} \epsilon(r) = 0$. Therefore,

$$
\begin{aligned}
|W(z) - W(y)| &= |V(T(z) - T(y))| \\
&\leq |V||T(z) - T(y)| \\
&\leq N|y - z||V|(2\delta + \epsilon(r))
\end{aligned}
$$

and so

$$|W(z)| \leq |V||h| + |W(z) - W(0)|.$$

Take $r$ such that

$$N\epsilon(r)|V| < \frac{1}{6}, \quad r \leq r_1,$$

and choose $\delta$ such that

$$\delta \leq \delta_1, \ |V||\delta| \leq \frac{r}{6}, \ 2N\delta|V| < \frac{1}{6}$$

and

$$\delta \leq \frac{1}{2}l(r) = \frac{1}{2}\min \left\{ |\phi(x) - p| : x \notin Q(a_1, r) \cup \dots \cup Q(a_k, r) \right\}.$$

We obtain

$$|W(z)| \leq r$$

and

$$|W(z) - W(y)| \leq \frac{|y - z|}{3},$$

for every $y, z \in \bar{Q}(0, r)$; hence $W : \bar{Q}(0, r) \to \bar{Q}(0, r)$ is a contraction and the equation $W(z) = z$ admits a unique solution in $\bar{Q}(0, r)$.
*Claim 2.* $\psi^{-1}(p) \subset Q(r)$.

Assume that $\psi(x) = p$ for some $x \in \bar{D} \setminus Q(r)$. Then

$$|\phi(x) - \psi(x)| \geq l(r) \geq 2\delta > |\phi(x) - \psi(x)|,$$

yielding a contradiction, and this proves the claim. We set

$$\psi^{-1}(p) = \{b_1, \dots, b_k\}.$$

Using claim 2 and the fact that $Q(r) \cap \partial D = \emptyset$, we conclude that $p \notin \psi(\partial D)$. We have

$$d(\phi, D, p) = \sum_{i=1}^k \text{sgn}(J_\phi(a_i))$$

and

$$d(\psi, D, p) = \sum_{i=1}^{k} \text{sgn}(J_\psi(b_i)).$$

Since for every $i = 1, \ldots, k$, $J_\phi$ is continuous in $Q(a_i, r)$ and $J_\phi(x) \neq 0$ for every $x \in Q(a_i, r)$, we deduce that $J_\phi$ has a constant sign in $Q(a_i, r)$ and so

$$\text{sgn}(J_\phi(a_i)) = \text{sgn}(J_\phi(b_i)).$$

Finally, as $|J_\phi(b_i) - J_\psi(b_i)| \leq \frac{c}{3}$ we have $\text{sgn}(J_\phi(b_i)) = \text{sgn}(J_\psi(b_i))$ and we conclude that

$$d(\phi, D, p) = d(\psi, D, p).$$

$\square$

The following result will play a crucial role in the extension of the notion of degree to the critical points $p \in \phi(Z_\phi)$.

**Lemma 1.4 [Sard's Lemma]** *Let $\phi \in C^1(\bar{D})^N$. Then $\phi(Z_\phi)$ is a set of measure zero.*

**Proof** Since $\phi \in C^1(\bar{D})^N$ and $\bar{D}$ is a compact set, there exists $M > 0$ such that

$$|\phi(x) - \phi(y)| + |\nabla\phi(x) - \nabla\phi(y)| \leq M|x - y| \tag{1.2}$$

for every $x, y \in \bar{D}$. For $x \in \bar{D}$ define $T_x : \bar{D} \to \mathbb{R}^N$ as

$$T_x(y) := \phi(x) + \nabla\phi(x)(y - x).$$

Using the fact that $\nabla\phi$ is uniformly continuous on $\bar{D}$, for every $\epsilon > 0$ we may choose $\delta > 0$ such that

$$\left| \frac{\partial\phi_i(x)}{\partial x_j} - \frac{\partial\phi_i(y)}{\partial x_j} \right| \leq \frac{\epsilon}{N}$$

for every $i, j = 1, \ldots, N$, and for every $x, y \in \bar{D}$ such that $|x - y| \leq \delta$. This implies that

$$|\phi(y) - T_x(y)| \leq \epsilon|x - y| \tag{1.3}$$

if $|x - y| \leq \delta$.

Assume without loss of generality that $D$ is a cube with side length $l > 0$ (or else, we write $D$ as a countable union of cubes $D_k$ and we prove that each $\phi(Z_\phi \cap D_k)$ is a set of zero measure). Fix $s \in \mathbb{N}$ large enough such that $2\frac{l}{s} < \delta$ and divide $D$ into $s^N$ cubes $D_k, k = 1, \ldots, s^N$, of side length $\frac{l}{s}$. First, we show that if some $D_k$ contains a critical point, then $\phi(D_k)$ has small measure. Indeed, if $x \in D_k$ and $J_\phi(x) = 0$, then for every $y \in D_k$ we have $|x - y| \leq \frac{l}{s} < \delta$ and, by (1.2), (1.3),

$$|\phi(x) - \phi(y)| \leq 2M\frac{l}{s}, \quad |\phi(y) - T_x(y)| \leq 2\epsilon\frac{l}{s}. \tag{1.4}$$

Since $J_\phi(x) = 0$, $T_x$ maps $\bar{D}$ into a hyperplane $P_x$ of dimension $N - 1$ and from (1.4) we obtain $\rho(\phi(y), P_x) \leq 2\epsilon\frac{l}{s}$. Therefore, $y \in D_k$ implies that $\phi(y)$ is in the cube which has $N - 1$ sides of length less than $4M\frac{l}{s}$ in $P_x$ and the $N$th side length less than $4\epsilon\frac{l}{s}$. Hence

$$\mathcal{L}^N(\phi(D_k)) \leq (4l)^N M^{N-1} \frac{\epsilon}{s^N}$$

and

$$\mathcal{L}^N(\phi(Z_\phi)) \leq (4l)^N M^{N-1} \epsilon.$$

Letting $\epsilon$ go to zero we obtain that $\mathcal{L}^N(\phi(Z_\phi)) = 0.$                    □

**Remark 1.5**

(i) The following generalization of Sard's Lemma can be found in de Rham (1955). Let $K$ and $L$ be two manifolds of dimension $k$ and $l$ respectively; let $f : K \to L$ be a function of class $r \geq \max\{k - l, 0\} + 1$. Then the set of critical values of $f$ has zero measure in $L$, where here critical points correspond to points where $\nabla f$ is not surjective.

(ii) In Chapter 4 we will prove a strong generalization of Sard's Lemma for Sobolev functions, namely if $p > N$ and if $f \in W^{1,p}(D)^N$, then for every measurable set $E \subset \bar{D}$, $f(E)$ is measurable and

$$\mathcal{L}^N(f(E)) \leq \int_E J_f(x)\, dx.$$

**Lemma 1.6** *Let $f \in C_c^1(\mathbb{R}^N)$, $K := \mathrm{spt}\,(f)$ and let $D$ be a domain of $\mathbb{R}^N$. Let $\gamma : [0,1] \to \mathbb{R}^N$ be a continuous path such that*

$$A := \{k + \gamma(s) \ : \ k \in K, s \in [0,1]\} \subset D. \tag{1.5}$$

*Then there exists a function $v \in C_c^1(D)^N$ such that*

$$\mathrm{div}\, v(x) = f(x - \gamma(0)) - f(x - \gamma(1)).$$

**Proof**

*Case 1.* Suppose first that $\gamma(s) \equiv s\bar{x}$ and set

$$F(x) := \int_0^1 f(x - \theta\bar{x})\, d\theta, \quad v(x) = \bar{x}F(x).$$

Clearly, $F \in C^1(\bar{D})$ and $\mathrm{spt}\,(F) \subset A$. Indeed, if $x \in \bar{D}$ is such that $F(x) \neq 0$, then there exists $\theta \in [0,1]$ such that $f(x - \theta\bar{x}) \neq 0$ and so $x - \theta\bar{x} \in K$, i.e. $x \in K + \theta\bar{x}$. Hence $x \in A$ and, as $A \subset\subset D$, we conclude that $v \in C_c^1(D)^N$. Moreover, we have

$$\mathrm{div}\, v(x) = \sum_{i=1}^N \bar{x}_i \int_0^1 \sum_{j=1}^N \frac{\partial f}{\partial x_j}(x - \theta\bar{x})\frac{\partial x_j}{\partial x_i}\, d\theta$$

$$= \sum_{i=1}^{N} \bar{x}_i \int_0^1 \frac{\partial f}{\partial x_i}(x - \theta\bar{x})\, d\theta$$

$$= -\int_0^1 \frac{d}{d\theta} f(x - \theta\bar{x})\, d\theta$$

$$= f(x) - f(x - \bar{x}).$$

*Case 2.* For a more general $\gamma$ satisfying (1.5), let $\mathcal{R}$ be the relation defined on $[0,1]$ by $t\mathcal{R}s$ if there exists $v \in C_c^1(D)^N$ such that

$$\text{div } v(x) = f(x - \gamma(s)) - f(x - \gamma(t)).$$

It is easy to see that $\mathcal{R}$ is an equivalence relation and its equivalence classes determine a partition of $[0,1]$. It suffices to show that each equivalence class is an open set of $[0,1]$ to conclude that $0\mathcal{R}1$.

Let $s \in [0,1]$ and let $C$ be the equivalence class for $s$. We must show that $C$ is an open set of $[0,1]$. Set

$$x_t := \gamma(t) - \gamma(s), t \in [0,1].$$

We claim that for a suitable $\epsilon > 0$ we have

$$\{t \in [0,1] \ : \ |t - s| < \epsilon\} \subset C.$$

Set $f_s(x) = f(x - \gamma(s)), K_s = \text{spt}\,(f_s)$ and $\eta = \frac{1}{2}\rho(K_s, D^c) > 0$. There exists an $\epsilon > 0$ such that $|x_t| < \eta$ whenever $|t - s| < \epsilon$. Fixing $t$ such that $|t - s| < \epsilon$ and setting

$$A_s := \{k + \theta x_t : \ k \in K_s, \ \theta \in [0,1]\},$$

we have $A_s \subset D$. Indeed, if we assume that for some $\theta \in [0,1]$, $x = k + \theta x_t \in D^c$, then we obtain $|x - k| = \theta|x_t| < \theta\eta$ and $\rho(k, D^c) < \eta = \frac{1}{2}\rho(K_s, D^c)$, which yields a contradiction. Finally, using case 1 we have $v \in C_c^1(D)^N$ such that

$$\text{div } v(x) = f_s(x) - f_s(x - x_t) = f(x - \gamma(s)) - f(x - \gamma(t)).$$

Hence $t\mathcal{R}s$, and $C$ is an open set of $[0,1]$. $\qquad\square$

The following representation formula was used by Heinz (1959) to introduce the topological degree.

**Proposition 1.7** *Let $\phi \in C^1(\bar{D})^N$, $p \notin \phi(\partial D) \cup \phi(Z_\phi)$, and let $f_\epsilon \in C_c^1(\mathbb{R}^N)$ be such that $\int_{\mathbb{R}^N} f_\epsilon(x)\, dx = 1$ and $\text{spt}\,(f_\epsilon) \subset Q(0, \epsilon)$. Then there exists $\epsilon(p) > 0$ such that*

$$d(\phi, D, p) = \int_D f_\epsilon(\phi(x) - p)J_\phi(x)\, dx$$

*for every $0 < \epsilon < \epsilon(p)$.*

**Proof** Since $p \notin \phi(Z_\phi)$, either $\phi^{-1}(p) = \emptyset$ or $\phi^{-1}(p) = \{a_1, \ldots, a_k\}$ (see Exercise 1.2).

First, suppose that $\phi^{-1}(p) = \emptyset$. By Definition 1.2 we have $d(\phi, D, p) = 0$. Setting $\delta := \rho(p, \phi(\bar{D})) > 0$ we have $|\phi(x) - p| \geq \delta > \epsilon$ for every $x \in \bar{D}$, $0 < \epsilon < \delta$. Therefore, $f_\epsilon(\phi(x) - p) = 0$ for every $x \in \bar{D}$ and so

$$\int_D f_\epsilon(\phi(x) - p) J_\phi(x) \, dx = 0.$$

Second, we assume that

$$\phi^{-1}(p) = \{a_1, \ldots, a_k\}.$$

Choose $r > 0$ such that $\bar{B}(a_i, r) \subset\subset D$, $i = 1, \ldots, k$, $\bar{B}(a_i, r) \cap \bar{B}(a_j, r) = \phi$ if $i \neq j$, and $J_\phi(x) \neq 0$ for every $x \in \cup_{i=1}^k \bar{B}(a_i, r)$. By the Inverse Function Theorem there exists $\epsilon_1 > 0$ such that

$$Q(p, \epsilon_1) \subset \phi(\bar{B}(a_i, r)), \quad \text{for every } i = 1, \ldots, k.$$

Also, there exists $\epsilon_2 > 0$ such that

$$|\phi(x) - p| < \epsilon_2, \quad x \in D \Rightarrow x \in \bigcup_{i=1}^k \bar{B}(a_i, r).$$

Hence, considering $\epsilon < \min\{\epsilon_1, \epsilon_2\}$

$$\int_D f_\epsilon(\phi(x) - p) J_\phi(x) \, dx = \int_{|\phi(x) - p| < \epsilon} f_\epsilon(\phi(x) - p) \, |J_\phi(x)| \operatorname{sgn}(J_\phi(x)) \, dx$$

$$= \sum_{i=1}^k \int_{\bar{B}(a_i, r) \cap \phi^{-1}(Q(p, \epsilon))} f_\epsilon(\phi(x) - p) \, |J_\phi(x)| \operatorname{sgn}(J_\phi(a_i)) \, dx$$

$$= \sum_{i=1}^k \operatorname{sgn}(J_\phi(a_i)) \int_{\phi(\bar{B}(a_i, r)) \cap Q(p, \epsilon)} f_\epsilon(z - p) \, dz$$

$$= \sum_{i=1}^k \operatorname{sgn}(J_\phi(a_i)) \int_{Q(0, \epsilon)} f_\epsilon(y) \, dy$$

$$= \sum_{i=1}^k \operatorname{sgn}(J_\phi(a_i)) = d(\phi, D, p).$$

$\square$

**Proposition 1.8** Let $\phi \in C^1(\bar{D})^N$, let $\Omega$ be a connected component of $\mathbb{R}^N \setminus \phi(\partial D)$, and let $p_1, p_2 \in \Omega \setminus \phi(Z_\phi)$. Then

$$d(\phi, D, p_1) = d(\phi, D, p_2).$$

**Proof**  Recall that if $\Omega$ is an open set of $\mathbb{R}$, then $\Omega$ is connected if and only if $\Omega$ is path-connected. Assume, in addition, that $\phi \in C^2(\bar{D})^N$. Let $\gamma : [0,1] \to \Omega$ be a continuous path such that $\gamma(0) = p_1$ and $\gamma(1) = p_2$. Let $f_\epsilon : \mathbb{R}^N \to \mathbb{R}$ be a family of continuous functions such that $\text{spt}(f_\epsilon) =: K_\epsilon \subset Q(0, \epsilon)$ and $\int_{\mathbb{R}^N} f_\epsilon(y)\,dy = 1$. By Proposition 1.7 there exists $\epsilon_0 > 0$ such that $0 < \epsilon \leq \epsilon_0$ implies that

$$d(\phi, D, p_i) = \int_D f_\epsilon(\phi(x) - p_i) J_\phi(x)\,dx, \quad i = 1, 2.$$

Take

$$\epsilon_1 := \frac{1}{2} \min\{\epsilon_0,\ \rho(\gamma, \Omega^c)\}$$

and set

$$A := \{k + \gamma(s)\ :\ k \in K_{\epsilon_1},\ s \in [0,1]\}.$$

It is clear that $A \subset \Omega$ and so by Lemma 1.6 there exists $v \in C_c^1(\Omega)^N$ such that

$$\text{div}\ v(x) = f_{\epsilon_1}(x - p_1) - f_{\epsilon_1}(x - p_2)$$

and $\text{spt}(v) \cap \phi(\partial D) \subset \Omega \cap \phi(\partial D) = \emptyset$. Hence (see Exercise 1.4) there exists $u \in C_c^1(D)^N$ such that

$$\text{div}\ u(x) = \text{div}\ v(\phi(x)) J_\phi(x) = [f_{\epsilon_1}(\phi(x) - p_1) - f_{\epsilon_1}(\phi(x) - p_2)] J_\phi(x),$$

and by Green's formula we conclude that

$$\begin{aligned}
d(\phi, D, p_1) - d(\phi, D, p_2) &= \int_D [f_{\epsilon_1}(\phi(x) - p_1) - f_{\epsilon_1}(\phi(x) - p_2)] J_\phi(x)\,dx \\
&= \int_D \text{div}\ v(\phi(x)) J_\phi(x)\,dx \\
&= \int_D \text{div}\ u(x)\,dx \\
&= 0.
\end{aligned}$$

Now we consider $\phi \in C^1(\bar{D})^N$. Let $\Omega$ and $\gamma : [0,1] \to \Omega$ be as in the first case. By Proposition 1.3 there exist $\delta(p_i) > 0, i = 1, 2$ such that $\|\psi - \phi\|_1 \leq \delta(p_i)$ implies $p_i \notin \psi(\partial D) \cup \psi(Z_\psi)$ and $d(\phi, D, p_i) = d(\psi, D, p_i)$. Let

$$\delta := \frac{1}{2} \min\{\delta(p_1), \delta(p_2), \rho(\gamma, \phi(\partial D))\}.$$

We show that $p_1, p_2$ are in the same connected component of $\mathbb{R}^N \setminus \psi(\partial D)$ provided that $\|\phi - \psi\| < \delta$.

For $x \in \partial D, s \in [0,1]$, we have

$$|\gamma(s) - \psi(x)| = |\gamma(s) - \phi(x) + \phi(x) - \psi(x)|$$

$$\geq \rho(\gamma(s), \phi(\partial D)) - \frac{1}{2}\rho(\gamma, \phi(\partial D))$$
$$\geq \frac{1}{2}\rho(\gamma, \phi(\partial D))$$
$$> 0.$$

Therefore, $\gamma(s) \in \mathbb{R}^N \setminus \psi(\partial D)$ for every $s \in [0,1]$ and since $\gamma$ joins $p_1$ and $p_2$, we deduce that $p_1$ and $p_2$ belong to the same connected component of $\mathbb{R}^N \setminus \psi(\partial D)$. Finally, taking $\psi \in C^2(\bar{D})^N$ such that $||\psi - \phi||_1 \leq \delta$, we have

$$d(\phi, D, p_1) = d(\psi, D, p_1) = d(\psi, D, p_2) = d(\phi, D, p_2).$$

$\square$

We now extend the notion of degree to critical values.

**Definition 1.9** Let $\phi \in C^1(\bar{D})^N$, $p \notin \phi(\partial D)$ such that $p \in \phi(Z_\phi)$. We define $d(\phi, D, p)$, the degree of $\phi$ at $p$ with respect to $D$, to be the number $d(\phi, D, q)$ for any $q \notin (\phi(Z_\phi) \cup \phi(\partial D))$ such that $|p - q| < \rho(p, \phi(\partial D))$.

*Justification*

(i) By Sard's Lemma we have $Q(p, r) \not\subset \phi(Z_\phi)$ for every $r > 0$. Therefore, there exists $q_r \in Q(p, r)$ such that $q_r \notin \phi(Z_\phi)$. Moreover, for $r$ small enough, $Q(p, r) \cap \phi(\partial D) = \emptyset$ since $\phi(\partial D)$ is a compact set and $p \notin \phi(\partial D)$. We conclude that there exists $q \notin (\phi(Z_\phi) \cup \phi(\partial D))$ such that $|p - q| < \rho(p, \phi(\partial D))$.

(ii) Assume that $|q_i - p| < \rho(p, \phi(\partial D))$, $i = 1, 2$ are such that $q_i \notin \phi(\partial D) \cap \phi(Z_\phi)$, $i = 1, 2$. Then $q_i \in B(p, \rho(p, \phi(\partial D))) \subset \mathbb{R}^N \setminus \phi(\partial D)$, $i = 1, 2$. Since $B(p, \rho(p, \phi(\partial D)))$ is a connected set which is included in $\mathbb{R}^N \setminus \phi(\partial D)$, we deduce that $B(p, \rho(p, \phi(\partial D)))$ is included in a connected component of $\mathbb{R}^N \setminus \phi(\partial D)$. Therefore, $q_1, q_2$ and $p$ are in the same connected component and, by Proposition 1.8,

$$d(\phi, D, q_1) = d(\phi, D, q_2).$$

**Remark 1.10**

(1) In general, if $\phi \in C^1(\bar{D})^N$ we cannot expect the inequality $|d(\phi, D, p)| \leq \sharp(\{\phi^{-1}(p)\})$ to hold for a critical value $p$. Indeed, identify $\mathbb{R}^2$ with $\mathbb{C}$ (the set of complex numbers), let $D$ be the unit ball of $\mathbb{C}$, let $n$ be an integer, and let $\phi \in C^1(\bar{D}, \mathbb{C})$ be defined by

$$\phi(z) := z^n, \quad z \in \bar{D} \subset \mathbb{C}.$$

We have $\phi^{-1}(0) = 0$ and $d(\phi, D, 0) = n$, hence

$$|d(\phi, D, 0)| > \sharp(\{\phi^{-1}(0)\}), \tag{1.6}$$

for every $n > 1$.

(2) By Sard's Lemma the set of points for which (1.6) holds has measure zero.

**Definition 1.11** *Let $\phi, \psi \in C^1(\bar{D})^N$ and $H : \bar{D} \times [0,1] \to \mathbb{R}^N$. We say that $H$ is a $C^1$ homotopy between $\phi$ and $\psi$ if*

(i) $H_t \in C^1(\bar{D})^N$ *for every $t \in [0,1]$,*
(ii) $\lim_{t \to s} \|H_t - H_s\|_1 = 0$ *for every $s \in [0,1]$,*
(iii) $H_0(x) = \phi(x)$, $H_1(x) = \psi(x)$ *for every $x \in \bar{D}$, where we set $H_t(x) = H(x,t), x \in \bar{D}, t \in [0,1]$.*

**Theorem 1.12** *Let $\phi \in C^1(\bar{D})^N$.*

(i) *$d(\phi, D, .)$ is constant on each connected component of $\mathbb{R}^N \setminus \phi(\partial D)$.*
(ii) *If $p \notin \phi(\partial D)$, then there exists $\epsilon > 0$ such that $\|\psi - \phi\|_1 \le \epsilon$ implies that $p \notin \psi(\partial D)$ and $d(\phi, D, p) = d(\psi, D, p)$.*
(iii) *If $H$ is a $C^1$ homotopy between $\phi$ and $\psi$ and $p \notin H_t(\partial D)$ for every $t \in [0,1]$, then $d(\phi, D, p) = d(\psi, D, p)$.*
(iv) *If $p \notin \phi(\partial D)$, then $d(\phi + a, D, p + a) = d(\phi, D, p)$ for every $a \in \mathbb{R}^N$.*

**Proof**

(i) Let $\Omega$ be a connected component of $\mathbb{R}^N \setminus \phi(\partial D)$ and $p_1, p_2 \in \Omega$. If $p_1 \notin \phi(Z_\phi)$ we set $q_1 = p_1$, and, if not, by Sard's Lemma we choose $q_1 \notin \phi(Z_\phi)$ such that $|p_1 - q_1| < \rho(p_1, \phi(\partial D))$. It is obvious that $q_1 \in \Omega$ since $q_1 \in B(p_1, \rho(p_1, \phi(\partial D))) \subset \mathbb{R}^N \setminus \phi(\partial D)$ and similarly, we select $q_2 \in \Omega$ such that $q_2 \in B(p_2, \rho(p_2, \phi, \partial D))$. By Proposition 1.8 $d(\phi, D, q_1) = d(\phi, D, q_2)$ and the result now follows from Definition 1.9.

(ii) By Sard's Lemma there exists $q \in \mathbb{R}^N \setminus \phi(\partial D)$ such that $q \notin \phi(Z_\phi)$ and $|q - p| < \frac{1}{2}\rho(p, \phi(\partial D))$. Hence $p$ and $q$ belong to the same component of $\mathbb{R}^N \setminus \phi(\partial D)$ and, by Proposition 1.3, there exists $0 < \epsilon_0 \equiv \epsilon_0(q, \phi) < \frac{1}{2}\rho(p, \phi(\partial D))$ such that

$$q \notin \psi(Z_\phi) \cup \psi(\partial D) \quad \text{and} \quad d(\psi, D, q) = d(\phi, D, q)$$

whenever $\|\phi - \psi\|_1 \le \epsilon_0$. For every $x \in \partial D$ we have

$$|\psi(x) - p| \ge |\phi(x) - p| - |\psi(x) - \phi(x)| > |p - q|,$$

therefore

$$|p - q| < \frac{1}{2}\rho(p, \phi(\partial D)) \le \rho(p, \psi(\partial D))$$

and so $p, q$ belong to the same connected component of $\mathbb{R}^N \setminus \psi(\partial D)$. By (i) we conclude that

$$d(\psi, D, p) = d(\psi, D, q) = d(\phi, D, q) = d(\phi, D, p),$$

where we have used Proposition 1.3 and Definition 1.9.

(iii) Define $u : [0,1] \to \mathbb{Z}$ as $u(t) = d(H_t, D, p)$. We show that $u$ is continuous. Fix $t \in [0,1]$. By (ii) there exists $\epsilon > 0$ such that $||H_t - H_s||_1 \leq \epsilon$ implies $d(H_t, D, p) = d(H_s, D, q)$. Since $\lim_{t \to s} ||H_t - H_s||_1 = 0$ there exists $\delta > 0$ such that $|t - s| < \delta$ implies $||H_t - H_s||_1 \leq \epsilon$, and so $d(H_t, D, p) = d(H_s, D, q)$. Thus $u$ is continuous on $[0,1]$. Since $[0,1]$ is a connected set and $u(t) \in \mathbb{Z}$ for every $t \in [0,1]$, we deduce that $u$ is a constant in $[0,1]$ hence $u(0) = u(1)$, i.e.

$$d(\phi, D, p) = d(\psi, D, p).$$

(iv) follows immediately from Proposition 1.7.

$\square$

An immediate consequence of Theorem 1.12 (iii) is that the topological degree depends uniquely on the boundary values. Precisely,

**Corollary 1.13** *Let $\phi, \psi \in C^1(\bar{D})$ be such that $\phi(x) = \psi(x)$ if $x \in \partial D$. Then for every $p \in \mathbb{R}^N \setminus \phi(\partial D)$,*

$$d(\phi, D, p) = d(\psi, D, p).$$

**Proof** Consider the convex homotopy

$$H(x,t) := t\phi(x) + (1 - t)\psi(x).$$

Clearly, $p \notin H(\partial D, t)$ for every $t \in [0,1]$ and the result follows from Theorem 1.12 (iii).                                                                            $\square$

**Remark 1.14** If $\phi \in C^1(\bar{D})$, $p \in \mathbb{R}^N \setminus \phi(\partial D)$, and if $\Omega$ is the connected component of $\mathbb{R}^N \setminus \phi(\partial D)$ containing $p$, then

$$\int_D f(\phi(x)) \det(\nabla\phi(x)) \, dx = d(\phi, D, p) \int_D f(y) \, dy$$

whenever $f \in C_c(\Omega)$.

The proof of this result is left as an exercise (see Exercise 1.5).

## 1.2   Topological degree for continuous functions

To introduce the definition of Brouwer degree for continuous functions, we use Tietze's Extension Theorem and we approximate continuous mappings by differentiable mappings.

**Theorem 1.15 [Tietze Extension Theorem]** *Let $X$ be a metric space, $A \subset X$ be a closed set, and let $f : A \to \mathbb{R}$ be a bounded, continuous function. Then there exists a continuous function $g : X \to \mathbb{R}$ which coincides with $f$ on $A$ such that*

$$\sup_{x \in X} g(x) = \sup_{x \in A} f(x) \quad and \quad \inf_{x \in X} g(x) = \inf_{x \in A} f(x).$$

We give a proof of a more general result in Theorem 7.26 in the case where $X$ is a normed space.

**Definition 1.16** *A function $\theta : \mathbb{R}^N \to \mathbb{R}$ is said to be a positive symmetric mollifier if*

   (i) $\theta \in C_c^\infty(\mathbb{R}^N)$,
   (ii) $\mathrm{spt}\,(\theta) \subset \{x \in \mathbb{R}^N : |x|_2 \le 1\}$,
   (iii) $\int_{\mathbb{R}^N} \theta(x)\,dx = 1$,
   (iv) $\theta(x) = \mu(|x|_2)$ *for some* $\mu : \mathbb{R}^+ \to \mathbb{R}$,
   (v) $\theta(x) \ge 0$ *for every* $x \in \mathbb{R}^N$.

As an example, the function $\theta : \mathbb{R}^N \to \mathbb{R}$ defined by

$$\theta(x) := \begin{cases} C \exp(\frac{1}{|x|_2^2 - 1}) & |x|_2 < 1 \\ 0 & |x|_2 \ge 1 \end{cases}$$

is a positive symmetric mollifier if the constant $C$ is chosen so that $\int_{\mathbb{R}^N} \theta(x)\,dx = 1$ (see Exercise 1.1).

**Lemma 1.17** *Let $D \subset \mathbb{R}^N$ be an open, bounded set and let $f : \bar{D} \to \mathbb{R}$ be a continuous function. Then for every $\epsilon > 0$ there exists $\tilde{f} \in C^\infty(\mathbb{R}^N)$ such that*

$$|\tilde{f}(x) - f(x)|_2 \le \epsilon.$$

*for every $x \in \bar{D}$.*

**Proof** Let $g : \mathbb{R}^N \to \mathbb{R}$ be a continuous extension of $f$ given by the Tietze Extension Theorem. Define $\theta_r : \mathbb{R}^N \to \mathbb{R}$ by $\theta_r(x) := \frac{1}{r^N}\theta(\frac{x}{r})$ and introduce the convolution

$$f_r := \theta_r * g.$$

Clearly, $f_r \in C^\infty(\mathbb{R}^N)$ for every $r > 0$. Since $g$ is uniformly continuous on the compact set $K = \{x \in \mathbb{R}^N : \rho(x, \bar{D}) \le 1\}$, there exists $0 < r < 1$ such that $y, z \in K$ and $|y - z|_2 \le r$ imply

$$|g(y) - g(z)| \le \frac{\epsilon}{\mathcal{L}^N(K)}.$$

For every $x \in \bar{D}$ we have

$$|f_r(x) - f(x)| = \left| \int_{\mathbb{R}^N} \theta_r(x - y)(g(y) - g(x))\,dy \right| \le \epsilon$$

and it suffices to set $\tilde{f} := f_r$. □

**Definition 1.18** *Let $\phi \in C(\bar{D})^N$ and $p \in \mathbb{R}^N \setminus \phi(\partial D)$. We define $d(\phi, D, p)$, the degree of $\phi$ at $p$ with respect to $D$, to be $d(\psi, D, p)$ for any $\psi \in C^1(\bar{D})^N$ such that $|\psi(x) - \phi(x)| < \rho(p, \phi(\partial D))$ for every $x \in \bar{D}$.*

*Justification.* Applying Lemma 1.17 to each component $\phi_i$, we may find $\psi \in C^1(\bar{D})^N$ such that $|\psi(x) - \phi(x)| < \rho(p, \phi(\partial D))$ for every $x \in \bar{D}$. Since $p \notin \phi(\partial D)$, $\psi \in C^1(\bar{D})^N$, and $|\psi(x) - \phi(x)| < \rho(p, \phi(\partial D))$ for every $x \in \bar{D}$, we have $p \notin \phi(\partial D)$.

Next we claim that, if $\psi_1, \psi_2 \in C^1(\bar{D})^N$ are such that $|\psi_1(x) - \phi(x)|, |\psi_2(x) - \phi(x)| < \rho(p, \phi(\partial D))$ for every $x \in \bar{D}$, then $d(\psi_1, D, p) = d(\psi_2, D, p)$. Indeed, let

$$H(x, t) := t\psi_1(x) + (1 - t)\psi_2(x), \quad x \in \bar{D}, \ t \in [1, 0].$$

$H$ is a $C^1$ homotopy between $\psi_1$ and $\psi_2$ and $p \notin H(\partial D, t)$ for every $t \in [0, 1]$ because for every $x \in \bar{D}$ we have

$$
\begin{aligned}
|H(x, t) - \phi(x)| &= |t(\psi_1(x) - \phi(x)) + (1 - t)(\psi_2(x) - \phi(x))| \\
&\leq t|(\psi_1(x) - \phi(x))| + (1 - t)|(\psi_2(x) - \phi(x))| \\
&< t\rho(p, \phi(\partial D)) + (1 - t)\rho(p, \phi(\partial D)) \\
&= \rho(p, \phi(\partial D)).
\end{aligned}
$$

By Theorem 1.12 we conclude that

$$d(\psi_1, D, p) = d(\psi_2, D, p).$$

**Proposition 1.19** *In Definition 1.18 the function $\psi$ can be chosen such that $p \notin \psi(Z_\psi)$.*

**Proof** Take $\chi \in C^1(\bar{D})^N$ such that

$$\|\phi - \chi\| < \frac{1}{2}\rho(p, \phi(\partial D))$$

and by Sard's Lemma choose $q \in \mathbb{R}^N$ such that $q \notin \chi(Z_\chi)$ and $|p - q| < \frac{1}{2}\rho(p, \phi(\partial D))$. Set

$$\psi(x) := \chi(x) + p - q.$$

Clearly, $\psi \in C^1(\bar{D})$ and

$$|\psi(x) - \phi(x)| \leq |\chi(x) - \phi(x)| + |p - q|$$

$$< \rho(p, \phi(\partial D)).$$

Also, one can see that $\psi(x) = p$ is equivalent to $\chi(x) = q$ and, for every $x \in \bar{D}$, $J_\psi(x) = J_\chi(x)$, hence $q \notin \chi(Z_\chi)$ implies that $p \notin \psi(Z_\psi)$.

It remains to prove that $p \notin \psi(\partial D)$. If $p \in \psi(\partial D)$, then $q \in \chi(\partial D)$ and so

$$\rho(p, \chi(\partial D)) \leq |p - q| + \rho(q, \chi(\partial D))$$

$$= |p - q|$$

$$< \frac{1}{2}\rho(p, \phi(\partial D)),$$

while

$$|\chi(x) - p| \geq |\phi(x) - p| - |\chi(x) - \phi(x)|$$

$$> \tfrac{1}{2}\rho(p, \phi(\partial D))$$

implies that

$$\rho(p, \chi(\partial D)) > \frac{1}{2}\rho(p, \phi(\partial D)),$$

yielding a contradiction.                                                                    □

We show that the degree does not change under composition with a diffeomorphism. We recall that $f : \mathbb{R}^N \to \mathbb{R}^N$ is a $C^1$ diffeomorphism if $f^{-1}$ exists and both $f$ and $f^{-1}$ are $C^1$.

**Theorem 1.20** Let $f : \mathbb{R}^N \to \mathbb{R}^N$ be a $C^1$ diffeomorphism and let $E \subset \mathbb{R}^N$ be a bounded, open set such that $f(\bar{E}) = \bar{D}$. Let $p \notin \phi(\partial D), q = f^{-1}(p)$, and let $\psi = f^{-1} \circ \phi \circ f$, where $\phi \in C(\bar{D}, \mathbb{R}^N)$. Then $d(\phi, D, p) = d(\psi, E, q)$.

**Proof** Since $f$ is a diffeomorphism we have $f(\partial E) = \partial D$ and so

$$q \in \psi(\partial E) \Leftrightarrow f^{-1}(p) \in f^{-1} \circ \phi \circ f(\partial E) \Leftrightarrow p \in \phi(\partial D),$$

which implies that $q \notin \psi(\partial E)$.
*Step 1.* Assume that $\phi \in C^1(D)^N$ and $p \notin \phi(Z_\phi)$. Clearly, $q \notin \psi(Z_\psi)$ and

$$
\begin{aligned}
d(\psi, E, q) &= \sum_{\psi(y)=q} \mathrm{sgn}(J_\psi(y)) \\
&= \sum_{(f^{-1}\circ\phi\circ f)(y)=f^{-1}(p)} \mathrm{sgn}\left[\det\left(\nabla\phi(f(y))\right)\left(\frac{\det\nabla f(y)}{\det\nabla f((f^{-1}\circ\phi\circ f)(y))}\right)\right] \\
&= \sum_{(\psi\circ f)(y)=p} \mathrm{sgndet}\left(\nabla\phi(f(y))\right) \\
&= \sum_{\phi(x)=p} \mathrm{sgndet}\left(\nabla\phi(x)\right) \\
&= d(\phi, D, p).
\end{aligned}
$$

*Step 2.* If $p \in \phi(Z_\phi)$, by Sard's Lemma we approximate $p$ by $\{p_n\} \subset \phi(Z_\phi)^c$; we obtain that $\{q_n\} = \{f^{-1}(p_n)\} \subset \phi(Z_\psi)^c$ and the result now follows from Step 1 and Theorem 1.12 (i).
*Step 3.* If $\phi \in C(\bar{D})^N$, we approximate $\phi$ by $\phi_n \in C^1(\bar{D})^N$ converging uniformly to $\phi$. Setting $\psi_n := f^{-1} \circ \phi_n \circ f$, then $\psi_n \in C^1(\bar{E})^N$, $\psi_n$ converges uniformly to $\psi$ and we conclude using Step 2 and Theorem 1.12 (ii).                                    □

**Remark 1.21** The theory developed in Section 1.2 is still valid if we replace $\mathbb{R}^N$ by an oriented space of finite dimension with a fixed basis. By Theorem 1.20 the definition of degree does not depend on the choice of fixed basis.

## 1.3   Generalization of the degree

In the previous sections we studied the degree for continuous functions $\phi : D \subset \mathbb{R}^N \to \mathbb{R}^N$. Here we indicate briefly how to construct the degree in a more general framework and, in particular, the following lemma indicates a natural way to introduce this concept for continuous functions $\phi : D \subset \mathbb{R}^N \to \mathbb{R}^M$, where $M \leq N$ are positive integers. In what follows, we identify $\mathbb{R}^M$ with $\{x \in \mathbb{R}^N : x_{M+1} = \ldots = x_N = 0\}$ and we consider $\phi : D \subset \mathbb{R}^N \to \mathbb{R}^M \subset \mathbb{R}^N$, where

$$\phi_{M+1} \equiv \ldots \equiv \phi_N \equiv 0.$$

We define $\psi \in C(\bar{D})^N$ by

$$\psi(x) := x + \phi(x)$$

and we set

$$\chi := \psi|_{D \cap \mathbb{R}^N}.$$

**Lemma 1.22** *If $\phi \in C(\bar{D})^N$ and if $p \in \mathbb{R}^M \setminus \psi(\partial D)$, then*

$$d(\psi, D, p) = d(\chi, D \cap \mathbb{R}^M, p).$$

**Proof**   We start by showing that $\psi^{-1}(p) = \chi^{-1}(p)$.

The inclusion $\chi^{-1}(p) \subset \psi^{-1}(p)$ is obvious. If $x \in \psi^{-1}(p)$, then $p = \psi(x) = x + \phi(x)$, which implies that $x = p - \phi(x) \in \mathbb{R}^M$. Therefore, $x \in \bar{D} \cap \mathbb{R}^M$ and since $p \notin \psi(\partial D)$ we have $x \in D \cap \mathbb{R}^M$ and so $p = \chi(x)$.

Next, we prove that if $D \cap \mathbb{R}^M = \emptyset$, then $d(\psi, D, p) = d(\chi, D^M, p)$. Indeed, $D \cap \mathbb{R}^M = \emptyset$, implies that $d(\chi, D \cap \mathbb{R}^M, p) = 0$, and we have

$$\emptyset = (D \cap \mathbb{R}^M) \cap \chi^{-1}(p) = D \cap \psi^{-1}(p) = \bar{D} \cup \psi^{-1}(p),$$

since $p \notin \psi(\partial D)$. Therefore, $p \notin \psi(\bar{D})$ and so

$$d(\psi, D, p) = 0.$$

*Step 1.* If $\phi \in C^1(\bar{D})^M, D \cap \mathbb{R}^M \neq \emptyset$, and if $p \notin \chi(Z_\chi)$, then $d(\psi, D, p) = d(\chi, D \cap \mathbb{R}^M, p)$.

Indeed, we have

$$\nabla\psi(x) = \begin{pmatrix} \nabla\chi(x) & B \\ 0_{M,N-M} & I_{N-M} \end{pmatrix},$$

where

$$B := \begin{pmatrix} \dfrac{\partial\phi_1}{\partial x_{M+1}} & \cdots & \dfrac{\partial\phi_1}{\partial x_N} \\ \vdots & & \vdots \\ \dfrac{\partial\phi_M}{\partial x_{M+1}} & \cdots & \dfrac{\partial\phi_M}{\partial x_N} \end{pmatrix}.$$

Therefore,

$$\mathrm{sgn}(J_\psi(x)) = \mathrm{sgn}(J_\chi(x))$$

for every $x \in D \cap \mathbb{R}^M$, which, together with the fact that $p \notin \chi(Z_\chi)$ and also $\psi^{-1}(p) = \chi^{-1}(p)$, yields $p \notin \psi(Z_\psi)$ and

$$d(\psi, D, p) = \sum_{x \in \psi^{-1}(p)} \mathrm{sgn}(J_\psi(x)) = \sum_{x \in \chi^{-1}(p)} \mathrm{sgn}(J_\chi(x)) = d(\chi, D \cap \mathbb{R}^M, p).$$

*Step 2.* If $\phi \in C(\bar{D})^M$, then $d(\psi, D, p) = d(\chi, D \cap \mathbb{R}^M, p)$.
　　Choose $\hat{\phi} \in C^1(\bar{D})^M$ such that

$$|\hat{\phi}(x) - \phi(x)| < \frac{1}{2}\rho(p, \psi(\partial D)), \tag{1.7}$$

for every $x \in \bar{D}$, set

$$\hat{\psi}(x) := x + \hat{\phi}(x), \quad x \in \bar{D},$$

and let $\hat{\chi}$ be the restriction of $\hat{\psi}$ to $\overline{D \cap \mathbb{R}^M}$. By Sard's Lemma there exists $a \in \mathbb{R}^M$ such that

$$|a| < \frac{1}{2}\rho(p, \psi(\partial D)), \quad p + a \notin \hat{\chi}(Z_{\hat{\chi}}).$$

By (1.7) we have

$$|\hat{\psi}(x) - \psi(x)| < \frac{1}{2}\rho(p, \psi(\partial D))$$

for every $x \in \bar{D}$ and so (see Definition 1.18)

$$d(\hat{\psi}, D, p) = d(\psi, D, p) \tag{1.8}$$

and, in the same way,

$$d(\hat{\chi}, D \cap \mathbb{R}^M, p) = d(\chi, D \cap \mathbb{R}^M, p). \tag{1.9}$$

By Step 1, Theorem 1.12 (iv) and as $p + a \notin \hat{\chi}(Z_{\hat{\chi}})$, we deduce that

$$d(\hat{\psi} + a, D, p + a) = d(\hat{\chi} + a, D \cap \mathbb{R}^M, p + a),$$

which implies

$$d(\hat{\psi}, D, p) = d(\hat{\chi}, D \cap \mathbb{R}^M, p).$$

This, together with (1.8) and (1.9), concludes the proof.　　　　□

　　In the remaining part of this section, we briefly introduce the notion of topological degree on manifolds and for more details we refer the reader to Nirenberg (1974) and Spivak (1979).

In what follows $\mathbb{M}$ is a $C^\infty$, paracompact, oriented manifold of dimension $N$; we denote by $d$ the operator of exterior differentiation on $\mathbb{M}$ and by $\wedge$ the exterior product operator between two differential forms on $\mathbb{M}$.

We recall that if $w$ is a smooth $(N-1)$-form on $\mathbb{M}$, then $w$ can be written as

$$w = \sum_{j=1}^{N} (-1)^{j-1} g_j(y) dy_1 \wedge \ldots \wedge dy_{j-1} \wedge dy_{j+1} \wedge \ldots \wedge dy_N$$

in local coordinates $(y_1, \ldots, y_N)$ and so

$$dw = \sum_{j=1}^{N} \frac{\partial g_j}{\partial y_j}(y) dy_1 \wedge \ldots \wedge dy_N.$$

Consider a $C^\infty$ $N$-form on $\mathbb{M}$,

$$\mu = f\, dy_1 \wedge \ldots \wedge dy_N$$

where $f : \mathbb{M} \to \mathbb{R}$ is a $C^\infty$ mapping.

Let $\bar{\Omega} \subset \mathbb{R}^N$ be an open, connected set and let $\phi : \Omega \to \mathbb{M}$ be a $C^\infty$ mapping. The *pull-back* of the form $\mu$ by $\phi$ is given by

$$\phi^*(\mu) := f(\phi(x))(\det \nabla \phi(x))\, dx_1 \wedge \ldots \wedge dx_N.$$

In general, when $\phi : \mathbb{P} \to \mathbb{M}$ is a $C^\infty$ mapping and $\mathbb{P}$ is a smooth $N$-dimensional manifold, then the pull-back $\phi^*(\mu)$ is an $N$-form on $\mathbb{P}$.

If $\operatorname{spt}(\mu) \subset \phi(\bar{\Omega})$ and if $\phi$ is an orientation-preserving diffeomorphism, then we define

$$\int_{\mathbb{M}} \mu := \int_{\Omega} \phi^*(\mu) = \int_{\Omega} f(\phi(x)) \det \nabla \phi(x)\, dx.$$

This definition does not depend on the choice of $\phi$ (see Spivak 1979, Chapter 8, Theorem 5). By Stoke's Theorem it follows that if $w$ is a smooth $(N-1)$-form with compact support on $\mathbb{M}$, then

$$\int_{\mathbb{M}} dw = 0.$$

**Lemma 1.23 [Sard's Lemma for Manifolds]** *Let $F : \mathbb{M} \to \mathbb{M}$ be a $C^1$ mapping. Then the set of critical values of $F$ has measure zero in $\mathbb{M}$.*

The proof is similar to that of Sard's Lemma in $\mathbb{R}^N$ (see Lemma 1.4), and it can be found in Nirenberg (1974) and Spivak (1979).

**Lemma 1.24** *Assume that $\mu$ is a $C^\infty$ $N$-form on $\mathbb{M}$ with $\int_{\mathbb{M}} \mu = 0$ and $\operatorname{spt} \mu \subset \Omega$, where $\Omega$ is a connected subset of $\mathbb{M}$. Then there exists an $(N-1)$-form $w$ on $\mathbb{M}$ such that $\operatorname{spt} w \subset \Omega$ and $\mu = dw$.*

For the proof of this result we refer the reader to Schwartz (1969). Motivated by Remark 1.14, we introduce the notion of degree of $\phi$ at $p \in$ M.

**Definition 1.25** *Let* M *be a* $C^\infty$*, paracompact, oriented manifold of dimension* $N$*, let* $D \subset$ M *be an open set with compact closure, let* $\phi \in C(\bar{D}; M) \cap C^1(D; M)$*, and* $p \in M \setminus \phi(\partial D)$*. Let* $\Omega \subset$ M *be the connected component of* $M \setminus \phi(\partial D)$ *containing* $p$ *and let* $\mu = f dy_1 \wedge \ldots \wedge dy_N$ *be a* $C^\infty$ $N$*-form with support contained in* $\Omega$ *such that* $\int_M \mu = 1$*. We say that* $\mu$ *is* admissible *for* $\phi$ *and* $p$ *and we define the degree of* $\phi$ *at* $p$ *with respect to* $D$ *by*

$$d(\phi, D, p) := \int_D \phi^*(\mu).$$

*Justification.* Let $\mu_1, \mu_2$ be two admissible $N$-forms and set $\mu := \mu_1 - \mu_2$. Then spt $\mu \subset \Omega$, $\int_M \mu = 0$, and, by Lemma 1.24, there exists an $(N-1)$-form $w$ such that spt $w \subset \Omega$ and $\mu = dw$. Hence,

$$\int_D \phi^*(\mu_1) - \int_D \phi^*(\mu_2) = \int_D \phi^*(\mu)$$

$$= \int_D \phi^*(dw)$$

$$= \int_D d(\phi^*(w))$$

$$= 0$$

by Stoke's Theorem, and so $d(\phi, \mu, p)$ is well defined.

It follows immediately from Definition 1.25 that if $p, q \in \Omega$, then

$$d(\phi, D, p) = d(\phi, D, q)$$

and also, by Exercise 1.6, $d(\phi, D, p)$ is an integer number. To extend the notion of degree to continuous mappings, we use the following facts:

1. $\lim_{q \to p} d(\phi, D, q) = d(\phi, D, p)$.
2. If $p$ is a regular value then $d(\phi, D, p) = \sum_{x \in \phi^{-1}(p)} \text{sgn}(J_\phi(x))$.
3. The degree is invariant under $C^{-1}$ homotopy.

**Definition 1.26** *Let* M *be a* $C^\infty$*, paracompact, oriented manifold of dimension* $N$*, let* $D \subset$ M *be an open set with compact closure, let* $\phi \in C(\bar{D}, M)$*, and* $p \in M \setminus \phi(\partial D)$*. Then*

$$d(\phi, D, p) := \lim_{n \to +\infty} d(\phi_n, D, p)$$

*where* $\{\phi_n\}_{n \in \mathbb{N}} \subset C(\bar{D}, M) \cap C^1(D; M)$ *is any sequence converging to* $\phi$ *uniformly on* $\bar{D}$*.*

We recall that $\bar{B}^N$ is the unit ball in $\mathbb{R}^N$, $S^{N-1}$ is the unit sphere, and $|x|_2$ stands for $\sqrt{x_1^2 + \ldots + x_n^2}$.

**Proposition 1.27** *Let $\phi \in C(\bar{B}^N)^N$ be such that $\phi(S^{N-1}) \subset \mathbb{R}^N \setminus \{0\}$. Let $\psi(x) := \frac{\phi(x)}{|\phi(x)|_2}$ for $x \in S^{N-1}$. Then*

$$d(\phi, B^N, 0) = d(\psi, S^{N-1}, p) =: d(\psi, S^{N-1}, S^{N-1})$$

*for every $p \in S^{N-1}$.*

**Proof** We notice first that $d(\phi, B^N, 0)$ is well defined since $0 \notin \phi(\partial B^N)$. Without loss of generality, assume that $\phi \in C^1(\bar{B}^N)^N$ and that $p$ is not a critical value (this can be done by virtue of Sard's Lemma in $\mathbb{R}^N$). Considering the homotopy

$$\phi_t(x) := \frac{\phi(x)}{|\phi(x)|_2^t}, \quad x \in \bar{B}, \ 0 \leq t < 1$$

and setting

$$G(x) := \begin{cases} |x|_2^2 \psi(\frac{x}{|x|_2}) & x \neq 0 \\ 0 & x = 0, \end{cases}$$

we observe that, for every $0 \leq t < 1$, $\phi_t(x) = 0$ whenever $\phi(x) = 0$ and so $\phi_t(x)$ is well defined everywhere and continuous in $\bar{B}$ and

$$d(\phi_t, B^N, 0) = d(\phi, B^N, 0).$$

As $\phi_t$ converges uniformly to $G$ on $S^{N-1}$ and as $G$ is continuous and $G(x) \neq 0$, for $x \in S^{N-1}$, there exists $t_0 \in (0, 1)$ such that the homotopy

$$sG(x) + (1 - s)\phi_{t_0}(x)$$

never vanishes for $x \in S^{N-1}$ and for $s \in [0, 1]$. Hence, by Theorem 1.12,

$$d(G, B^N, 0) = d(\phi_{t_0}, B^N, 0) = d(\phi, B^N, 0).$$

Since $p$ is not a critical value we have $\psi^{-1}(p)$ finite, i.e.

$$\{x_1, \ldots, x_k\} = \psi^{-1}(p),$$

and we may find $\epsilon_0 > 0$ small enough such that $p_0 := \epsilon^2 p$ is a regular value of $\phi$ for all $0 < \epsilon < \epsilon_0$. We have

$$G(a) = p_0 \Leftrightarrow a \in \{\epsilon x_1, \ldots, \epsilon x_k\};$$

hence, by Exercise 1.6,

$$d(\psi, S^{N-1}, p) = \sum_{i=1}^{k} \mathrm{sgn}(J_\psi(x_k))$$

and

$$d(G, B^N, \epsilon^2 p) = \sum_{i=1}^{k} \operatorname{sgn}(J_G(\epsilon x_i)).$$

Noticing that $\operatorname{sgn}(J_\psi(x_i)) = \operatorname{sgn}(J_G(\epsilon x_i))$, we conclude that

$$d(\phi, B^N, 0) = d(G, B^N, 0) = d(\psi, S^{N-1}, S^{N-1}),$$

where we have used the fact that, for $\epsilon$ small,

$$d(G, B^N, \epsilon^2 p) = d(G, B^N, 0).$$

This property can easily be deduced from Proposition 1.8 within the framework of continuous functions. This will be proved in detail in Theorem 2.3. □

### 1.4 Exercises

**Exercise 1.1** Let $\theta : \mathbb{R}^N \to \mathbb{R}$ be defined by

$$\theta(x) = \begin{cases} \exp\left(\frac{1}{|x|_2^2 - 1}\right) & |x|_2 < 1 \\ 0 & |x|_2 \geq 1. \end{cases}$$

Prove that $\theta$ is integrable in $\mathbb{R}^N$ and $\theta \in C_c^\infty(\mathbb{R}^N)$.

*Solution 1.1.* Let $\mu : \mathbb{R} \to \mathbb{R}$ be defined by

$$\mu(t) := \begin{cases} \exp(\frac{1}{t^2 - 1}) & |t| < 1 \\ 0 & |t| \geq 1. \end{cases}$$

It is obvious that $\mu$ is infinitely differentiable at every point $t \neq -1, 1$. Note that for every $n \in \mathbb{N}$ there exists a polynomial $P_n(t)$ such that, for every $|t| < 1$,

$$\mu^{(n)}(t) = \frac{P_n(t)}{(t^2 - 1)^{2n}} \exp\left(\frac{1}{t^2 - 1}\right)$$

and so

$$\lim_{t \to 1} \mu^{(n)}(t) = \lim_{t \to -1} \mu^{(n)}(t) = 0.$$

Since $\mu^{(n)}(t) = 0$ for every $|t| > 1$ we deduce that $\mu^{(n)}(-1)$ and $\mu^{(n)}(1)$ exist and so $\mu \in C^\infty(\mathbb{R})$. Considering, in addition, the fact that the function $x \to |x|_2$ is infinitely differentiable, we deduce that $\theta \in C^\infty(\mathbb{R}^N)$. We also have

$$\operatorname{spt}(\theta) \subset \{x \in \mathbb{R}^N : |x|_2 \leq 1\}$$

and so $\theta$ is integrable on $\mathbb{R}^N$.

**Exercise 1.2** Let $D \subset \mathbb{R}^N$ be an open, bounded set, $\phi \in C^1(\bar{D}, \mathbb{R}^N)$, and $p \notin \phi(Z_\phi)$. Prove that $\phi^{-1}(p)$ is finite.

*Solution 1.2.* Assume that there exists a sequence $\{x_k\} \subset \phi^{-1}(p)$ such that $x_k \neq x_l$ for every $k \neq l$. Since $\bar{D}$ is a compact set we can assume, without loss of generality, that there exists $\bar{x} \in \mathbb{R}^N$ such that $x_k \to \bar{x}$. We obtain $\bar{x} \in \bar{D}$ and by the continuity of $\phi$, $\bar{x} \in \phi^{-1}(p)$. Using the fact that $\phi$ is continuously differentiable we have

$$0 = \phi(x_k) - \phi(\bar{x}) = \nabla\phi(\bar{x})(x_k - \bar{x}) + (x_k - \bar{x})\epsilon(x_k - \bar{x}), \qquad (1.10)$$

where $\epsilon$ is a real function satisfying $\lim_{t\to 0} \epsilon(t) = 0$. Since $\bar{x} \in \phi^{-1}(p)$ and $p \notin \phi(Z_\phi)$, we have

$$0 < \gamma := \inf\{|\nabla\phi(\bar{x})u| : u \in \mathbb{R}^N, |u| = 1\}. \qquad (1.11)$$

For $k$ large enough, from (1.10) and (1.11) we have

$$\gamma \leq \left|\nabla\phi(\bar{x})\left(\frac{x_k - \bar{x}}{|x_k - \bar{x}|}\right)\right| \leq \frac{\gamma}{2}.$$

This yields a contradiction, hence $\phi^{-1}(p)$ is finite.

**Exercise 1.3** Let $D \subset \mathbb{R}^N$ be an open, bounded set, $\phi \in C^2(\bar{D})^N$, and let $(A_{ij}(x))^1$ be the adjugate matrix of $\nabla\phi$. Prove that

$$\sum_{j=1}^{N} \frac{\partial}{\partial x_j} A_{ij}(x) = 0$$

for every $x \in \bar{D}$ and every $i = 1, \ldots, N$.

*Hint.* Using the formula

$$\det(C_1, \ldots, C_N) = \sum_{\sigma \in S_N} \text{sign}\,\sigma\, C_1^{\sigma(1)} \ldots C_N^{\sigma(N)},$$

where $S_N$ is the set of $N$ permutations and $\text{sign}\,\sigma$ is the signature of $\sigma$, prove first that if $C_i \in C^1(\bar{D})^N$, $i = 1, \ldots, N$, then

$$\frac{\partial}{\partial x_i}\det(C_1, \ldots, C_N) = \det\left(\frac{\partial C_1}{\partial x_i}, \ldots, C_N\right) + \ldots + \det\left(C_1, \ldots, \frac{\partial C_N}{\partial x_i}\right). \qquad (1.12)$$

Second, using (1.12) show that

$$\frac{\partial A_{ij}}{\partial x_j} = (-1)^{i+j}\sum_{p<i}(-1)^{p+1}\det\left(\frac{\partial^2\psi}{\partial x_p\partial x_i}, \frac{\partial\psi}{\partial x_1}, \ldots, \frac{\hat{\partial\psi}}{\partial x_p}, \ldots, \frac{\hat{\partial\psi}}{\partial x_i}, \ldots, \frac{\partial\psi}{\partial x_N}\right)$$

$^1 A(x)^T\nabla\phi = I_N\det\nabla\phi(x)$ where $I_N$ is the $n \times N$ identity matrix.

$$+ \sum_{p>i}(-1)^p \det\left(\frac{\partial^2\psi}{\partial x_p \partial x_i}, \frac{\partial\psi}{\partial x_1}, \ldots, \frac{\hat{\partial\psi}}{\partial x_i}, \ldots, \frac{\hat{\partial\psi}}{\partial x_p}, \ldots, \frac{\partial\psi}{\partial x_N}\right),$$

where $\hat{C}_i$ indicates that the column vector $C_i$ has been removed and $\psi :=(\phi_1, \ldots, \hat{\phi}_j, \ldots, \phi_N)$.

**Exercise 1.4** Let $D \subset \mathbb{R}^N$ be an open, bounded set, $\phi \in C^2(D)^N$, and let $v \in C_c^1(\mathbb{R}^N)^N$ be such that spt $(v) \cap \phi(\partial D) = \emptyset$. Prove that there exists $u \in C_c^1(\mathbb{R}^N)^N$ such that

$$\operatorname{div} u(x) = J_\phi(x) \operatorname{div} v(\phi(x)).$$

*Solution 1.4.* Let $(A_{ij}(x))$ be the matrix of the cofactors of $\phi(x)$, as introduced in Exercise 1.3; set

$$u_i(x) := \sum_{j=1}^N v_j(\phi(x))A_{ji}(x).$$

We claim that spt $(u_i) \subset \phi^{-1}(\operatorname{spt}(v)) \cap \bar{D} \subset D$. Indeed, if $x \in \bar{D}$ is such that $u_i(x) \neq 0$, then there exists $j$ such that $v_j(\phi(x)) \neq 0$. Therefore, $\phi(x) \in \operatorname{spt}(v_j) \subset \operatorname{spt}(v)$, and $x \in \phi^{-1}(\operatorname{spt}(v))$. Moreover, $\phi^{-1}(\operatorname{spt}(v))$ is a compact set and so $\bar{D} \cap \phi^{-1}(\operatorname{spt}(v)) \subset \bar{D}$ is a compact set. Since $\phi^{-1}(\operatorname{spt}(v)) \cap \partial D = \phi$, we have $\bar{D} \cap \phi^{-1}(\operatorname{spt}(v)) \subset D$ and we conclude that spt $(u_i) \subset D$. It is obvious that $u \in C^1(\bar{D})^N$, hence $u \in C_0^1(\bar{D})^N$.

We show that $\operatorname{div} u(x) = J_\phi(x) \operatorname{div} v(\phi(x))$. We have

$$\frac{\partial u_i}{\partial x_i}(x) = \sum_{j=1}^N \left[\frac{\partial A_{ji}}{\partial x_i}(x)v_j(\phi(x)) + A_{ji}(x)\sum_{k=1}^N \frac{\partial v_j}{\partial x_k}(\phi(x))\frac{\partial \phi_k}{\partial x_i}(x)\right].$$

By Exercise 1.3,

$$\sum_{i,j}\frac{\partial A_{ji}}{\partial x_i}(x)v_j(\phi(x)) = \sum_j v_j(\phi(x))\sum_i \frac{\partial A_{ji}}{\partial x_i}(x) = 0$$

and

$$\sum_{i,j,k=1}^N A_{ji}(x)\frac{\partial v_j}{\partial x_k}(\phi(x))\frac{\partial \phi_k}{\partial x_i}(x) = \sum_{j,k=1}^N \frac{\partial v_j}{\partial x_k}(\phi(x))\delta_{j,k}J_\phi(x)$$

$$= J_\phi(x)\operatorname{div} v(\phi(x)).$$

Hence,

$$\operatorname{div} u(x) = J_\phi(x)\operatorname{div} v(\phi(x)).$$

**Exercise 1.5** Prove that if $\phi \in C^1(\bar{D})$, $p \in \mathbb{R}^N \backslash \phi(\partial D)$, and if $\Omega$ is the connected component of $\mathbb{R}^N \backslash \phi(\partial D)$ containing $p$, then

$$\int_D f(\phi(x)) \det (\nabla \phi(x)) \, dx = d(\phi, D, p) \int_D f(y) \, dy$$

whenever $f \in C_c(\Omega)$.

*Hint.* Without loss of generality assume that $f \geq 0$ (simply write $f = f^+ - f^-$). Note that

$$\int_D f(\phi(x)) \det (\nabla \phi(x)) \, dx = \int_U f(\phi(x)) \det (\nabla \phi(x)) \, dx,$$

where $U$ is the open set where $\det (\nabla \phi(x)) \neq 0$ and decompose $U$ into a union of disjoint open sets $U_i$ such that $\det (\nabla \phi(x))$ has a constant sign on each $U_i$, and a set of zero measure. Apply the Inverse Function Theorem in each $U_i$ and, using the definition of the degree for noncritical values, deduce the result (see also Theorem 5.30 for a more general proof).

**Exercise 1.6** Let $\mathbb{M}$ be $C^\infty$ paracompact, oriented manifolds of dimension $N$, let $D \subset \mathbb{M}$ be an open set with compact closure, let $\phi \in C(\bar{D}; \mathbb{M}) \cap C^1(D; \mathbb{M})$, and $p \in \mathbb{M} \backslash (\phi(\partial D) \cup \phi(Z_\phi))$. Prove that $d(\phi, D, p) = \sum_{x \in \phi^{-1}(p)} \mathrm{sgn}(J_\phi(x))$.

*Solution 1.6* (see also Spivak 1979, Chapter 8, Theorem 2). Assume first that $p \notin \phi(\bar{D})$, and let $\Omega$ be the connected component of $\mathbb{M} \backslash \phi(\partial D)$. Let $W$ be an open set contained in a coordinate patch of $p$ and let $\mu$ be a $C^\infty$ $N$-form with support contained in $W$ such that

$$\int_{\mathbb{M}} \mu = 1.$$

Then $\mu = f dy_1 \wedge \ldots \wedge dy_N$ and $\mathrm{spt}\, f \cap \phi(\bar{D}) = \emptyset$; hence

$$0 = \int_D \phi^*(\mu) = d(\phi, D, p).$$

Next, we assume that $p \in \phi(\bar{D})$. Since $p \notin \phi(Z_\phi)$, we know that $\phi^{-1}(p)$ must be finite,

$$\phi^{-1}(p) = \{x_1, \ldots, x_k\},$$

and we choose disjoint open sets $N_i \subset D$, $i = 1, \ldots, k$, such that $x_i \in N_i$ and $\phi|_{N_i}$ is a diffeomorphism. Set

$$V := \bigcap_{i=1}^k \phi(N_i).$$

Then $V$ is an open neighbourhood of $p$ and we choose a $C^\infty$ $N$-form on $\mathbb{M}$ with support contained in $V$ such that

$$\int_M \mu = 1.$$

We have

$$d(\phi, D, p) = \int_D \phi^*(\mu)$$

$$= \sum_{i=1}^k \int_{N_i} \phi^*(\mu)$$

$$= \sum_{i=1}^k \text{sgn}(J_\phi(x_i)) \int_{\phi(N_i)} \mu$$

$$= \sum_{i=1}^k \text{sgn}(J_\phi(x_i)).$$

# 2

# DEGREE THEORY IN FINITE-DIMENSIONAL SPACES

In this chapter we prove those properties of $d(\phi, D, p)$ that we consider to be most useful in applications.

## 2.1 Dependence of the degree on $\phi$ and $p$

The first theorem is the basis of many of the applications of degree theory in analysis.

**Theorem 2.1** *Let* $\phi \in C(\bar{D})^N$, $p \notin \phi(\partial D)$ *such that* $d(\phi, D, p) \neq 0$. *Then there exists* $x \in D$ *such that* $\phi(x) = p$.

**Proof** Assume that $p \notin \phi(D)$. Since $p \notin \phi(\partial D)$ we have $p \notin \phi(\bar{D})$, and as $\phi(\bar{D})$ is a compact set we have $\rho(p, \phi(\bar{D})) > 0$. By Proposition 1.19, choose $\psi \in C^1(\bar{D})^N$ such that $||\psi - \phi|| < \rho(p, \phi(\bar{D}))$ and $p \notin \psi(Z_\psi) \cup \psi(\partial D)$. We have $p \notin \psi(\bar{D})$ and then, using Definition 1.18,

$$0 = d(\psi, D, p) = d(\phi, D, p),$$

which yields a contradiction. $\qquad \square$

**Definition 2.2** $H : \bar{D} \times [0, 1] \to \mathbb{R}^N$ *is a* $C^0$ *homotopy between* $\phi, \psi \in C(\bar{D})^N$ *if* $H$ *is continuous on* $\bar{D} \times [0, 1]$, $H(x, 0) = \phi(x)$, *and* $H(x, 1) = \psi(x)$ *for every* $x \in \bar{D}$.

**Theorem 2.3** *Let* $\phi \in C(\bar{D})^N$ *and let* $p \notin \phi(\partial D)$. *Then*

(1) *for every* $\psi \in C(\bar{D})^N$ *such that* $||\psi - \phi|| < \rho(p, \phi(\partial D))$, *we have*

$$d(\phi, D, p) = d(\psi, D, p);$$

(2) *if* $H(x, t) =: h_t(x)$ *is a* $C^0$ *homotopy between* $h_0$, $h_1$ *and* $p \notin h_t(\partial D)$ *for every* $t \in [0, 1]$, *then* $d(h_t, D, p)$ *does not depend on* $t \in [0, 1]$;

(3) *if* $p_1, p_2$ *belong to the same connected component of* $\mathbb{R}^N \setminus \phi(\partial D)$, *then*

$$d(\phi, D, p_1) = d(\phi, D, p_2).$$

**Proof**

(1) Let $\chi \in C^1(\bar{D})^N$ be such that

$$||\psi - \chi|| < \rho(p, \phi(\partial D)) - ||\phi - \psi||.$$

Then

$$||\phi - \chi|| \leq ||\phi - \psi|| + ||\psi - \chi|| < \rho(p, \phi(\partial D)),$$

and so

$$d(\phi, D, p) = d(\chi, D, p).$$

Also, $||\psi - \chi|| < \rho(p, \psi(\partial D))$, because if $x \in \partial D$, then

$$||\psi(x) - p|| \geq ||\phi(x) - p|| - ||\phi(x) - \psi(x)|| > \rho(p, \phi(\partial D)) - ||\psi - \phi|| \geq ||\psi - \chi||.$$

Thus $d(\psi, D, p) = d(\chi, D, p)$ and we conclude that

$$d(\phi, D, p) = d(\psi, D, p).$$

(2) Let $u(t) = d(h_t, D, p)$. By (1) and the fact that $h_t$ is a $C^0$ homotopy, we obtain that $u$ is continuous on $[0, 1]$. Since $u(t) \in \mathbb{Z}$ for every $t \in [0, 1]$ we deduce that $u$ is constant on $[0, 1]$.

(3) Let $\Omega$ be the connected component of $\mathbb{R}^N \setminus \phi(\partial D)$ containing $p_1$ and $p_2$ and let $\gamma$ be a continuous path in $\Omega$ joining $p_1$ to $p_2$. Take $\psi \in C^1(\bar{D})^N$ such that $||\psi - \phi|| < \rho(\gamma, \Omega^c)$. We claim that

$$d(\psi, D, p_1) = d(\phi, D, p_1) \text{ and } d(\psi, D, p_2) = d(\phi, D, p_2).$$

Indeed, since $p_1 \in \gamma$ and $\phi(\partial D) \subset \Omega^c$, we obtain that

$$||\psi - \phi|| < \rho(p_1, \phi(\partial D))$$

and, by Definition 1.18,

$$d(\psi, D, p_1) = d(\phi, D, p_1).$$

By a similar argument we obtain that

$$d(\psi, D, p_2) = d(\phi, D, p_2).$$

It remains to show that $d(\psi, D, p_1) = d(\psi, D, p_2)$. Since

$$\rho(\gamma, \psi(\partial D)) > \rho(\gamma, \phi(\partial D)) - \rho(\phi(\partial D), \psi(\partial D))$$
$$> \rho(\gamma, \Omega^c) - \rho(\phi(\partial D), \psi(\partial D)) > 0,$$

we have that $\gamma \subset \mathbb{R}^N \setminus \psi(\partial D)$ and so $p_1, p_2$ are in the same connected component of $\mathbb{R}^N \setminus \psi(\partial D)$. By Proposition 1.8 we deduce that

$$d(\psi, D, p_1) = d(\psi, D, p_2).$$

$\square$

**Theorem 2.4** *Let* $\phi, \psi, \in C(\bar{D})^N$ *be such that* $\phi|_{\partial D} = \psi|_{\partial D}$. *Then* $d(\phi, D, p) = d(\psi, D, p)$ *for every* $p \notin \phi(\partial D)$.

**Proof** As $\phi(\partial D) = \psi(\partial D)$, $d(\phi, D, p)$ and $d(\psi, D, p)$ exist for every $p \notin \phi(\partial D)$. Set

$$H(x, t) := t\phi(x) + (1 - t)\psi(x), \quad x \in \bar{D}, \ t \in [0, 1].$$

$H$ is a $C^0$ homotopy between $\psi$ and $\phi$ and $H(\partial D, t) = \phi(\partial D)$. By Theorem 2.3 (2), $d(H(\cdot, t), D, p)$ does not depend on $t \in [0, 1]$ and so

$$d(\phi, D, p) = d(\psi, D, p).$$

$\square$

The next result will help to generalize the invariance of the degree under homotopy.

**Proposition 2.5** *Let* $\phi \in C(\bar{D})^N$, $p \in \phi(\partial D)$, *and* $q \in \mathbb{R}^N$. *Then*

$$d(\phi - q, D, p - q) = d(\phi, D, p).$$

**Proof** This follows immediately from the definition. $\square$

**Theorem 2.6** *Let* $h_t : \bar{D} \to \mathbb{R}^N$ *be a* $C^0$ *homotopy and let* $t \mapsto p_t$ *be a continuous function from* $[0, 1]$ *into* $\mathbb{R}^N$ *such that* $p_t \notin h_t(\partial D)$ *for every* $t \in [0, 1]$. *Then* $d(h_t, D, p_t)$ *is independent of* $t \in [0, 1]$.

**Proof** Set $k_t(x) := h_t(x) - p_t$. We have $p_t \notin h_t(\partial D)$ if and only if $0 \notin k_t(\partial D)$ and, by Proposition 2.5, for $t \in [0, 1]$ we have

$$d(k_t, D, 0) = d(h_t, D, p_t).$$

Since $k_t$ is a $C^0$ homotopy, by Theorem 2.3 we conclude that $d(k_t, D, 0)$ is independent of $t \in [0, 1]$ and so is $d(h_t, D, p_t)$. $\square$

## 2.2    Dependence of the degree on the domain $D$

**Theorem 2.7** *Let* $\phi \in C(\bar{D})^N$ *and let* $p \notin \phi(\partial D)$.

(1) (*Domain decomposition property*) *If* $D = \cup_{i \in \mathbb{N}} D_i$ *and* $D_i$ *are open, mutually disjoint sets, then*

$$d(\phi, D, p) = \sum_{i \in \mathbb{N}} d(\phi, D_i, p).$$

(2) (*Excision property*) *If* $K \subset \bar{D}$ *is a compact set such that* $p \notin \phi(K)$, *then*

$$d(\phi, D, p) = d(\phi, D \setminus K, p).$$

## Proof

(1) As $\partial D_j \subset \partial D$, $d(\phi, D_i, p)$ is well defined. Let $\psi \in C^1(\bar{D})^N$ be such that $||\phi - \psi|| < \rho(p, \phi(\partial D_i))$ and $p \notin \psi(Z_\psi)$ (see Proposition 1.19). We have

$$||\phi - \psi||_{D_i} < \rho(p, \phi(\partial D)) \le \rho(p, \phi(\partial D_i))$$

and so

$$d(\phi, D_i, p) = d(\phi, D_i, p).$$

Since $p \notin \psi(Z_\psi)$ and the sets $D_i$ are mutually disjoint, by Theorem 2.1 there are only finitely many $i$ such that $d(\psi, D_i, p) \ne 0$. Assume, without loss of generality, that $d(\psi, D_i, p) \ne 0$ for $i = 1, \dots, l$ and $d(\psi, D_i, p) = 0$ for every $i \ge l + 1$. We obtain

$$
\begin{aligned}
d(\phi, D, p) &= \sum_{x \in \psi^{-1}(p)} \mathrm{sgn}(J_\psi(x)) \\
&= \sum_{i=1}^{l} \sum_{x \in \psi^{-1}(p) \cap D_i} \mathrm{sgn}(J_\psi(x)) \\
&= \sum_{i=1}^{l} d(\psi, D_i, p) \\
&= \sum_{i=1}^{+\infty} d(\psi, D_i, p) \\
&= \sum_{i=1}^{+\infty} d(\phi, D_i, p).
\end{aligned}
$$

(2) Let $\psi \in C^1(\bar{D})^N$ be such that $||\phi - \psi|| < \rho(p, \phi(K \cup \partial D))$ and $p \notin \psi(Z_\psi)$. It is clear that $p \notin \psi(K)$ and so

$$d(\psi, D, p) = d(\psi, D \setminus K, p).$$

Note that $\partial(D \setminus K) \subset \partial D \cup K$. As $||\phi - \psi||_{(D \setminus K)} < \rho(p, \phi(\partial D \cup K)) \le \rho(p, \phi(\partial D))$ we have

$$d(\phi, D, p) = d(\psi, D, p).$$

Also, $||\phi - \psi||_{(D \setminus K)} \le ||\phi - \psi||_D < \rho(p, \phi(\partial D \cup K))$ and so

$$d(\phi, D \setminus K, p) = d(\psi, D \setminus K, p) = d(\psi, D, p) = d(\phi, D, p).$$

$\square$

Now we introduce the notion of the index of a $p$-point $x_0 \in D$.

**Definition 2.8** Let $\phi \in C(\bar{D})^N$, $p \notin \phi(\partial D)$, and let $x_0 \in D$ be an isolated $p$-point of $\phi$ (i.e. $\phi(x_0) = p$ and there exists a neighbourhood $V$ of $x_0$ such that

$V \cap \phi^{-1}(x_0) = \emptyset)$. *Let $\mathcal{U}$ be the collection of all open neighbourhoods $V$ of $x_0$ such that $\bar{V}$ does not contain another p-point of $\phi$. We define the* index *of $\phi$ with respect to $(x_0, p)$ by*

$$i(\phi, x_0, p) := d(\phi, V, p)$$

*for any $V \in \mathcal{U}$.*

*Justification*

(1) For $V \in \mathcal{U}$ we obtain that $p \notin \phi(\partial V)$ and so $d(\phi, V, p)$ is well defined.
(2) Let $V_1, V_2 \in \mathcal{U}$. We have $V = V_1 \cup V_2 \in \mathcal{U}$ and setting $K = \bar{V}_1 \cap V_2^c$ we have that $K \subset \bar{V}$ is a compact set and $p \notin \phi(K)$. By the excision property of the degree (see Theorem 2.7) we obtain that

$$\begin{aligned} d(\phi, V, p) &= d(\phi, V \setminus K, p) \\ &= d(\phi, V \cap (\bar{V}_1^c \cup V_2), p) \\ &= d(\phi, V_2, p). \end{aligned}$$

Using a similar argument, we have $d(\phi, V_1 \cup V_2, p) = d(\phi, V_1, p)$; thus

$$d(\phi, V_1, p) = d(\phi, V_2, p).$$

**Theorem 2.9** *Let $\phi \in C(\bar{D})^N$ and let $p \notin \phi(\partial D)$.*

(1) *If $\phi^{-1}(p)$ is finite, then $d(\phi, D, p) = \sum_{x \in \phi^{-1}(p)} i(\phi, x, p)$.*
(2) *If $\phi \in C^1(\bar{D})^N$, $a \in \phi^{-1}(p)$, and if $J_\phi(a) \neq 0$, then $a$ is an isolated point and $i(\phi, a, p) = (-1)^\nu$, where $\nu$ is the number of real negative eigenvalues of $\nabla\phi(a)$, counting algebraic multiplicity.*

*Warning.* In general we do not have $|i(\phi, x, p)| \leq 1$. As an example, consider

$$\phi(x, y) := (x^2 - y^2, 2xy),$$

where $(x, y) \in D := (-1, 1) \times (-1, 1)$. We have $d\left(\phi, D, \begin{pmatrix} 0 \\ 0 \end{pmatrix}\right) = 2$, $\phi^{-1}\begin{pmatrix} 0 \\ 0 \end{pmatrix} = \begin{pmatrix} 0 \\ 0 \end{pmatrix}$ and $i\left(\phi, \begin{pmatrix} 0 \\ 0 \end{pmatrix}, \begin{pmatrix} 0 \\ 0 \end{pmatrix}\right) = 2$.

**Proof of Theorem 2.9**

(1) Since $\phi^{-1}(p)$ is finite, then every $a \in \phi^{-1}(p)$ is an isolated point and so $i(\phi, a, p)$ is well defined. Assume that

$$\phi^{-1}(p) = \{a_1, \dots, a_k\}.$$

Let $V_1, \dots, V_K$ be open sets in $D$, mutually disjoint, and such that $a_1 \in V_1$, $\dots$, $a_k \in V_K$. By Definition 2.8 we obtain that $i(\phi, a_i, p) = d(\phi, V_i, p)$.

Therefore, using the excision property of the degree (see Theorem 2.7), we have

$$\sum_{x \in \phi^{-1}(p)} i(\phi, x, p) = \sum_{i=1}^{k} i(\phi, a_i, p)$$

$$= \sum_{i=1}^{k} d(\phi, V_i, p)$$

$$= d(\phi, \overset{k}{\underset{i=1}{\cup}} V_i, p)$$

$$= d(\phi, D \setminus K, p)$$

$$= d(\phi, D, p),$$

where $K = \bar{D} \setminus \cup_{i=1}^{k} V_i$.

(2) Let $V \subset D$ be an open set such that $a \in V$, $\phi(a) = p$ and $\phi(x) \neq p$ for every $x \in \bar{V}$, $x \neq a$. By Definition 2.8 we have

$$i(\phi, a_i, p) = d(\phi, V, p) = \mathrm{sgn}(J_\phi(a)).$$

Let $\lambda_1, \ldots, \lambda_n$ be the eigenvalues of $\nabla \phi(a)$. Then

$$J_\phi(a) = \lambda_1 \ldots \lambda_n.$$

where the complex eigenvalues occur in conjugate pairs $\alpha, \bar{\alpha}$ such that $\alpha \bar{\alpha} > 0$, and so

$$\mathrm{sgn}(J_\phi(a)) = (-1)^\nu.$$

$\square$

## 2.3   The multiplication theorem

Given two mappings $\phi \in C^1(\bar{D})^N$ and $\psi \in C^1(\phi(\bar{D}))^N$ we want to compute $d(\psi \circ \phi, D, p)$ in terms of $d(\psi, \Delta_i, p)$ and $d(\phi, D, \Delta_i)$, where $\Delta_i$ $(i \in \mathbb{N})$ are the connected components of $\mathbb{R}^N \setminus \phi(\partial D)$. This is obtained in the multiplication theorem.

**Theorem 2.10 [Multiplication Theorem]** *Let $\phi \in C(\bar{D})^N$, let $M \subset \mathbb{R}^N$ be an open set containing $\phi(\bar{D})$, $\Delta = M \setminus \phi(\partial D)$, and let $\psi \in C(\bar{M})^N$. Let $(\Delta_i)_{i \in \mathbb{N}}$ be the connected components of $\Delta$ and $p \notin \psi \circ \phi(\partial D) \cup \psi(\partial M)$. Then we have*

(1) $p \notin \psi(\partial \Delta_i)$ *for every $i \in \mathbb{N}$.*

(2) $d(\psi \circ \phi, D, p) = \sum_{i \in \mathbb{N}} d(\psi, \Delta_i, p) d(\phi, D, \Delta_i).$

**Proof** Note that $\partial \Delta_i \subset \partial M \cup \phi(\partial D)$ for every $i \in \mathbb{N}$. Indeed, if $q \in \partial \Delta_i$ for some $i \in \mathbb{N}$, then $q \in \partial \Delta_i \subset \overline{\partial \Delta_i} \subset \bar{M}$. Assume that $q \notin \partial M$. We obtain that $q \in M = \cup_{j \in \mathbb{N}} \Delta_j$, thus for a suitable $j \in \mathbb{N}$ we have $q \in \Delta_j$ and $i \neq j$, contradicting $\Delta_i \cap \Delta_j = \emptyset$. Hence

$$\partial \Delta_i \subset \partial M \cup \phi(\partial D) \qquad (2.1)$$

and $d(\varphi, \Delta_i, p)$ is well defined. Also, $p \notin \psi \circ \phi(\partial D)$ and so $d(\psi \circ \phi, D, p)$ is well defined. Since $\Delta_i$ is a connected component of $M \setminus \phi(\partial D)$, we deduce that $\Delta_i$ is a connected set included in $\mathbb{R}^N \setminus \phi(\partial D)$ and so $\Delta_i$ is a subset of a connected component of $\mathbb{R}^N \setminus \phi(\partial D)$, $D_i$. As $d(\phi, D, q)$ is a constant for $q \in D_i$, we denote this constant by $d(\phi, D, \Delta_i)$. We claim that there is only a finite number of $i$ such that $d(\psi, \Delta_i, p) \neq 0$. Indeed, take $f \in C^1(\bar{M})^N$ such that $||f - \psi|| < \rho(p, \psi(\partial M))$ and $p \notin f(Z_f)$. Since there are only finitely many $x$ such that $f(x) = p$ and since the sets $\Delta_i$ are mutually disjoint, there are finitely many $i$ so that $f(x) = p$ admits a solution in $\Delta_i$, say $\Delta_1, \ldots, \Delta_k$. We have $d(f, \Delta_i, p) = 0$ for every $i > k$. As

$$||f - \psi||_{\Delta_i} \leq ||f - \psi|| < \rho(p, \phi(\partial M)) \leq \rho(p, \phi(\partial \Delta_i)),$$

we have $d(f, \Delta_i, p) = d(\psi, \Delta_i, p)$ for every $i \in \mathbb{N}$ and so

$$d(\psi, \Delta_i, p) = 0$$

for every $i > k$.

The proof of

$$d(\psi \circ \phi, D, p) = \sum_i d(\psi, \Delta_i, p) d(\phi, D, \Delta_i)$$

is divided into four cases.

*Case 1.* We assume that $\psi \in C^1(\bar{M})^N$, $\phi \in C^1(\bar{D})^N$, and $p \notin \psi \circ \phi(Z_{\psi \circ \phi})$.
   Then $p \notin \psi(Z_\psi)$ and $y \notin \phi(Z_\phi)$ for every $y \in \psi^{-1}(p)$, hence

$$
\begin{aligned}
d(\psi \circ \phi, D, p) &= \sum_{x \in D,\, \psi\phi(x)=p} \mathrm{sgn}(J_{\psi \circ \phi}(x)) \\
&= \sum_{x \in D,\, \psi\phi(x)=p} \mathrm{sgn}(J_\psi(\phi(x)))\, \mathrm{sgn}(J_\phi(x)) \\
&= \sum_{y \in \Delta,\, \psi(y)=p} \sum_{x \in D,\, \phi(x)=y} \mathrm{sgn}(J_\psi(y))\, \mathrm{sgn}(J_\phi(x)) \\
&= \sum_{y \in \Delta,\, \psi(y)=p} \mathrm{sgn}(J_\psi(y))\, d(\phi, D, y) \\
&= \sum_{\Delta_i} \sum_{y \in \Delta_i,\, \psi(y)=p} \mathrm{sgn}(J_\psi(y))\, d(\phi, D, \Delta_i) \\
&= \sum_i d(\psi, \Delta_i, p) d(\phi, D, \Delta_i).
\end{aligned}
$$

*Case 2.* We assume that $\psi \in C^1(\bar{M})^N$, $\phi \in C^1(\bar{D})^N$, and $p \in \psi \circ \phi(Z_{\psi \circ \phi})$.

By Sard's Lemma there exists $q \notin \psi \circ \phi(Z_{\psi \circ \phi})$ such that

$$|p - q| < \min\{\rho(p, \psi \circ \phi(\partial D)), \ \rho(p, \psi(\partial M))\}.$$

By (2.1) $\partial \Delta_i \subset \partial M \cup \phi(\partial D)$ for every $i \in \mathbb{N}$; hence

$$|p - q| < \rho(p, \psi(\partial \Delta_i))$$

and so $p$ and $q$ belong to the same connected component of $\mathbb{R}^N \setminus \psi(\partial \Delta_i)$. This yields $d(\psi, \Delta_i, p) = d(\psi, \Delta_i, q)$ for every $i \in \mathbb{N}$ and so, by Case 1,

$$\begin{aligned}
d(\psi \circ \phi, D, p) &= d(\psi \circ \phi, D, q) \\
&= \sum_i d(\psi, \Delta_i, q) d(\phi, D, \Delta_i) \\
&= \sum_i d(\psi, \Delta_i, p) d(\phi, D, \Delta_i).
\end{aligned}$$

*Case 3.* We assume that $\psi \in C^1(\bar{M})^N$ and $\phi \in C(\bar{D})^N$.

Set

$$L := \psi^{-1}(p), \quad \epsilon := \frac{1}{4}\rho(L, \phi(\partial D)) \tag{2.2}$$

and let $\hat{\phi} \in C^1(\bar{D})^N$ be such that, for every $x \in \bar{D}$,

$$|\hat{\phi}(x) - \phi(x)| < \min\{\delta, \epsilon\}, \tag{2.3}$$

where $\delta > 0$ is such that

$$|a - b| \leq \delta \Rightarrow |\psi(a) - \psi(b)| < \rho(p, \psi \circ \phi(\partial D)). \tag{2.4}$$

Suppose, in addition, that $\hat{\phi}(\bar{D}) \subset M$. We have

$$d(\psi \circ \phi, D, p) = d(\psi \circ \hat{\phi}, D, p). \tag{2.5}$$

Let $\hat{\Delta} = M \setminus \hat{\phi}(\partial D)$ and let $\{\hat{\Delta}_j, j \in \mathbb{N}\}$ be the family of connected components of $\hat{\Delta}$. Set

$$A = \{\Delta_i \ : \ \Delta_i \cap L \neq \emptyset, \ d(\phi, D, \Delta_i) \neq 0\}$$

and

$$\hat{A} = \{\hat{\Delta}_i \ : \ \hat{\Delta}_i \cap L \neq \emptyset, \ d(\hat{\phi}, D, \hat{\Delta}_i) \neq 0\}.$$

Since $L$ is a compact set, both $A$ and $\hat{A}$ have finite cardinality. Write

$$A = \{\Delta_1, \ldots, \Delta_k\}, \text{ and } \hat{A} = \{\hat{\Delta}_1, \ldots, \hat{\Delta}_l\}.$$

We claim that $k = l$. Indeed, fix $i = 1, \ldots, k$, and $y \in \Delta_i \cap L$. By (2.2) and (2.3) we obtain that $|\hat{\phi}(x) - \phi(x)| < \rho(y, \phi(\partial D))$ for every $x \in \bar{D}$ and so

$$d(\hat{\phi}, D, y) = d(\phi, D, y) \neq 0. \tag{2.6}$$

We want to prove that $Q(y, \epsilon) \subset \mathbb{R}^N \setminus \hat{\phi}(\partial D)$. Assume, on the contrary, that there exists $x \in \partial D$ such that $|\hat{\phi}(x) - y| < \epsilon$. We obtain

$$|\phi(x) - y| \leq |\hat{\phi}(x) - y| + |\hat{\phi}(x) - \phi(x)| < 2\epsilon < \rho(y, \phi(\partial D)),$$

which yields a contradiction. Hence $Q(y, \epsilon) \subset \mathbb{R}^N \setminus \hat{\phi}(\partial D)$ and since $Q(y, \epsilon)$ is a connected set we deduce that $Q(y, \epsilon) \subset \hat{\Delta}_{j(i)}$ for some $j(i) \in \mathbb{N}$, and this implies that

$$k \geq l. \tag{2.7}$$

Using (2.6) and the fact that degree $d(\hat{\phi}, D, \cdot)$ is a constant on $\hat{\Delta}_{j(i)}$ we have

$$d(\phi, D, \Delta_i) = d(\hat{\phi}, D, \Delta_{j(i)}). \tag{2.8}$$

Note also that, by (2.2) and (2.3),

$$\epsilon \leq \frac{1}{3}\rho(y, \hat{\phi}(\partial D)). \tag{2.9}$$

Fix $j = 1, \ldots, l$, and $y \in \hat{\Delta}_i \cap L$. By (2.3) and (2.9), we have $|\hat{\phi}(x) - \phi(x)| < \epsilon < \rho(y, \hat{\phi}(\partial D))$ and so

$$d(\phi, D, y) = d(\hat{\phi}, D, y) \neq 0.$$

Using a similar argument, we prove that there exist $i(j) \in \mathbb{N}$ such that $Q(y, \epsilon) \subset \Delta_{i(j)}$ and

$$d(\phi, D, \Delta_i) = d(\hat{\phi}, D, \Delta_{j(i)}). \tag{2.10}$$

Hence

$$l \geq k,$$

which, together with (2.7), yields

$$l = k. \tag{2.11}$$

We show that $d(\psi, \Delta_i, p) = d(\psi, \hat{\Delta}_i, p)$ for every $i \in \mathbb{N}$. Using the excision property of degree (see Theorem 2.7), with $K = (\hat{\Delta} \setminus \hat{\Delta}_i) \cup (\bar{\Delta}_i \setminus \hat{\Delta}_i)$, we have

$$d(\psi, \hat{\Delta}_i, p) = d(\psi, \hat{\Delta}_i \cap \Delta_i, p) = d(\psi, \Delta_i, p). \tag{2.12}$$

By (2.2), (2.4), (2.5), (2.8), (2.10), (2.11), (2.12), and the decomposition formula for $\psi$ and $\hat{\phi}$, we deduce that

$$\begin{aligned} d(\psi \circ \phi, D, p) &= d(\psi \circ \hat{\phi}, D, p) \\ &= \sum_j d(\psi, \hat{\Delta}_j, p) d(\hat{\phi}, D, \hat{\Delta}_j) \end{aligned}$$

$$= \sum_i d(\psi, \Delta_i, p) d(\phi, D, \Delta_i).$$

*Case 4.* We assume that $\psi \in C(\bar{M})^N$ and $\phi \in C(\bar{D})^N$.

Let $\hat{\psi} \in C^1(\bar{M})^N$ be such that for every $y \in \bar{M}$

$$|\hat{\psi}(y) - \psi(y)| < \min\{\rho(p, \psi \circ \phi(\partial D)), \ \rho(p, \psi(\partial M))\}.$$

For every $x \in \bar{D}$ we have

$$|\hat{\psi} \circ \phi(x) - \psi \circ \phi(x)| < \min\{\rho(p, \psi \circ \phi(\partial D)), \ \rho(p, \psi(\partial M))\}$$

and by Proposition 1.19 and the decomposition formula for $\hat{\psi}$, $\phi$, obtained in Case 3, we have

$$d(\psi \circ \phi, D, p) = \sum_i d(\psi, \Delta_i, p) d(\phi, D, \Delta_i).$$

$\square$

## 2.4   An application of Hopf's theorem

In this section we state the well-known Hopf's Theorem. The idea is the following: we have already proved that the degree is invariant under a $C^0$ homotopy (see Theorem 2.3) and the question now is whether functions with the same degree must be homotopic. Hopf proved that this is true for balls (see Hopf 1926), i.e. if two continuous mappings defined on the unit ball $B(0,1) = B^N$ have the same degree at the origin, then they are $C^0$ homotopic.

**Theorem 2.11 [First Version of Hopf's Theorem]** *Let $N \geq 2$ and let $\Omega = B^N$ be the unit ball of centre $0$ in $\mathbb{R}^N$. Assume that $\phi, \psi \in C(\bar{\Omega})^N$, $0 \notin \phi(\partial\Omega) \cup \psi(\partial\Omega)$, and $d(\phi, \Omega, 0) = d(\psi, \Omega, 0)$. Then there exists a continuous function $H : [0,1] \times \bar{\Omega} \to \mathbb{R}^N$ such that $0 \notin H(t, \partial\Omega)$ for every $t \in [0,1]$ and $H(0, \cdot) = \phi$, $H(1, \cdot) = \psi$.*

We refer the reader to Guillemin and Pollack (1974) for an analytic proof of this result.

We state a second version of Hopf's Theorem which can be easily deduced from the first version.

**Theorem 2.12 [Second Version of Hopf's Theorem]** *Let $N \geq 2$ and let $S^{N-1} = \partial B^N$ be the unit sphere in $\mathbb{R}^N$. Assume that $\varphi_1, \varphi_2 \in C(S^{N-1}, S^{N-1})$ and $d(\varphi_1, S^{N-1}, S^{N-1}) = d(\varphi_2, S^{N-1}, S^{N-1})$. Then there exists a continuous function $H : S^{N-1} \times [0,1] \to S^{N-1}$ such that $H(\cdot, 0) = \varphi_1$, $H(\cdot, 1) = \varphi_2$.*

**Corollary 2.13** *Let $\varphi : S^{N-1} \to S^{N-1}$ be a continuous mapping. Then the following assertions are equivalent:*

   (i) $\varphi$ *is not homotopic to a constant.*

(ii) *Every continuous extension* $\phi : \bar{B}^N \to \mathbb{R}^N$ *admits a zero.*

(iii) *Every continuous extension* $\phi : \bar{B}^N \to \mathbb{R}^N$ *verifies* $d(\phi, B^N, 0) \neq 0.$

## Proof

*Step 1.* We start by showing that if $c \in S^{N-1}$, then $d(c, S^{N-1}, S^{N-1}) = 0$. In fact, by Proposition 1.27,

$$d(c, S^{N-1}, S^{N-1}) = d(c, B^N, 0)$$

and by Theorem 2.1 $d(c, B^N, 0) = 0$.

*Step 2.* Let

$$\mathcal{F} := \{\phi \in C(\bar{B}^N; \mathbb{R}^N) : \phi|_{S^{N-1}} = \varphi\}.$$

First, we prove that (i) $\Rightarrow$ (ii). By Hopf's Theorem, we have

$$d(\varphi, S^{N-1}, S^{N-1}) \neq 0.$$

For every $\phi \in \mathcal{F}$, by Proposition 1.27 we have

$$d(\phi, B^N, 0) = d(\varphi, S^{N-1}, S^{N-1}) \neq 0$$

and by Theorem 2.1 we deduce that there exists $x \in \bar{B}^N$ such that $\phi(x) = 0$.

Next, we show that (ii) $\Rightarrow$ (iii). Assume that $\varphi$ admits a continuous extension $\phi_1 : \bar{B} \to \mathbb{R}^N$ such that $d(\phi_1, B^N, 0) = 0$. Then by Proposition 1.27 and by Hopf's Theorem $\varphi$ is homotopic to a constant, i.e. given a constant $c \in S^{N-1}$, there exists a continuous mapping $H : S^{N-1} \times [0, 1] \to S^{N-1}$ such that, for every $x \in S^{N-1}$,

$$H(x, 0) = c, \quad H(x, 1) = \varphi(x).$$

Define $\phi : \bar{B}^N \to S^{N-1}$ by

$$\phi(x) := H\left(\frac{x}{|x|_2}, |x|_2\right), \quad x \in B^N.$$

It is clear that $\phi$ is continuous at every $x \in \bar{B}^N \backslash \{0\}$. Using the uniform continuity of $H$ and the fact that $H(x, 0) = c$ for every $x \in \bar{B}^N$, we deduce that $\phi$ is continuous at 0. Therefore, we find $\phi \in \mathcal{F}$ such that the equation $\phi(x) = 0$ has no solution in $B^N$.

Finally, we prove that (iii) $\Rightarrow$ (i). Assume that $\varphi$ is homotopic to a constant. As before, setting

$$\phi(x) := H\left(\frac{x}{|x|_2}, |x|_2\right), \quad x \in B^N,$$

it is easy to verify that $\phi$ is a continuous extension of $\varphi$, homotopic to a constant. By Theorem 2.3 and Step 1 we conclude that $d(\phi, B^N, 0) = 0$. $\qquad\square$

## 2.5    Degree and winding number

In this section, we establish the relation between the degree and the winding number. In order to achieve this goal, we recall some definitions and properties of holomorphic functions. Throughout this section $\mathbb{C}$ denotes the set of complex numbers (which we identify with $\mathbb{R}^2$) $i^2 = -1$, $D$ is an open, bounded, subset of $\mathbb{C}$, and the derivative $\phi'$ of a function $\phi : D \to \mathbb{C}$ is given by

$$\phi'(z) = \lim_{h \in \mathbb{C}, h \to 0} \frac{\phi(z+h) - \phi(z)}{h}.$$

If $\phi'(z)$ exists for all $z \in D$, then $\phi$ is said to be *holomorphic in* $D$, and we write $\phi \in H(D)$.

**Definition 2.14** *Assume that $\partial D$ is a $C^1$ closed curve, let $\phi : \bar{D} \to \mathbb{C}$ be a holomorphic function and let $p \notin \phi(\partial D)$. We define the* winding number *of $\phi$ at $p$ with respect to $\partial D$ by*

$$w(\phi, D, p) := \frac{1}{2i\pi} \int_{\partial D} \frac{\phi'(z)}{\phi(z) - p} \, dz.$$

**Lemma 2.15** *Let $\phi : D \to \mathbb{C}$ be a holomorphic function, $Z := \{z \in D \,|\, \phi(z) = 0\}$ and let $E$ be the set of limit points of $Z$. Then $E$ is both an open and a closed subset of $D$.*

**Remark 2.16**

(i) Assume that $D$ is a connected set. We recall that $\phi(z) = p$ then has an infinite number of solutions if and only if $\phi$ is a constant $p$.

(ii) It is easy to prove that $F := \{z \in D : \phi^{(n)}(z) = 0 \text{ for all } n \in \mathbb{N}\}$ is both closed and open in $D$. Indeed,

$$F = \bigcap_{n \in \mathbb{N} \cup \{0\}} \left\{ z \in D : \phi^{(n)}(z) = 0 \right\}$$

is a closed subset of $D$. Moreover, if $z_0 \in F$ by the analyticity of $\phi$ there exists $R > 0$ such that

$$\phi(z) = \sum_{n=0}^{+\infty} \frac{\phi^{(n)}(z_0)}{n!} (z - z_0)^n$$

for every $z \in B(z_0, R)$, and so

$$\phi(z) = \phi'(z) = \phi^{(2)}(z) = \ldots = 0$$

for every $z \in B(z_0, R)$. Hence,

$$B(z_0, R) \subset F$$

and $F$ is an open set.

**Lemma 2.17 [Cauchy's Lemma]** *Let $D$ be a convex set, $z_0 \in D$, and let $\phi \in H(D \setminus \{z_0\}) \cap C(\bar{D})$. Then there is a holomorphic function $F \in H(D)$ such that*

$$F' = \phi.$$

For the proof we refer the reader to Rudin (1966).

**Lemma 2.18** *Assume that $\phi \in H(D)$ is not constant, $z_0 \in D$, and $w_0 = \phi(z_0)$. Then*

(i) *there exists $m \in \mathbb{N}$ such that $z_0$ is a zero of $\phi - w_0$ of order $m$, i.e. $\phi(z) - w_0 = (z - z_0)^m g(z)$, where $g$ is holomorphic and $g(z_0) \neq 0$;*
(ii) *there is an open neighbourhood $V$ of $z_0$ and a function $\varphi \in H(V)$ such that*

$$\phi(z) = w_0 + (\varphi(z))^m$$

*and $\varphi$ is a bijection from $V$ into $B(0, r)$ for a suitable $r > 0$.*

**Proof** We may assume without loss of generality that

$$D = B(z_0, R_1) \text{ and } \phi(z) = \sum_{n=0}^{+\infty} a_n (z - z_0)^n$$

for every $z \in B(z_0, R_1)$.

(i) Let $\phi(z_0) = w_0$. Since $\phi$ is not constant, there is $n$ so that $a_n \neq 0$. Let $m \in \mathbb{N}$ be such that

$$a_1 = \ldots = a_{m-1} = 0, \ a_m \neq 0.$$

We obtain

$$\phi(z) = (z - z_0)^m \left[ \sum_{n=0}^{+\infty} a_{m+n} (z - z_0)^n \right]$$

and setting

$$g(x) := \sum_{n=0}^{+\infty} a_{m+n} (z - z_0)^n$$

we have

$$\phi(z) - w_0 = (z - z_0)^m g(z), \ g \in H(B(z_0, R_1)) \text{ and } g(z_0) \neq 0.$$

(ii) Let $0 < R_2 \leq R_1$ be such that

$$g(z) \neq 0$$

for every $z \in B(z_0, R_2)$. Then $\frac{g'(z)}{g(z)}$ is holomorphic in $B(z_0, R_2)$ and, by Cauchy's Lemma, there exists $h \in H(B(z_0, R_2))$ such that

$$h' = \frac{g'}{g}.$$

We have

$$(g \cdot \exp(-h))' = g' \cdot \exp(-h) - h' g \cdot \exp(-h) = g \cdot \exp(-h) \left( \frac{g'}{g}(1 - h') \right) = 0$$

and since $B(z_0, R_2)$ is a connected set, we deduce that there exists a constant $c \in \mathbb{C}$ such that

$$g \cdot \exp(-h) = c$$

for every $z \in B(z_0, R_2)$. Set

$$\varphi(z) := (z - z_0)c^{\frac{1}{m}} \exp \left( \frac{h}{m} \right).$$

We obtain that $\varphi \in H(B(z_0, R_2))$ and

$$(\varphi(z))^m = (z - z_0)^m c \cdot \exp(h) = g(z)(z - z_0)^m$$

and so

$$\phi(z) = w_0 + (\varphi(z))^m$$

for every $z \in B(z_0, R_2)$. Hence $\varphi(z_0) = 0$ and, in addition,

$$\varphi'(z) = c^{\frac{1}{m}} \exp \left( \frac{h}{m} \right) \left( 1 + \frac{z - z_0}{m} h'(z) \right)$$

thus $\varphi'(z_0) = c^{\frac{1}{m}} \exp \left( \frac{h(z_0)}{m} \right) \neq 0$. By the Inverse Function Theorem we deduce that there is an open set $V$ containing $z_0$ such that

$$\varphi : V \to B(0, r)$$

is a bijection for some $r > 0$.

□

**Lemma 2.19** *Let $\phi \in H(D)$ be such that*

$$\phi(x + iy) = \phi_1(x, y) + i\phi_2(x, y)$$

*for every $x, y \in \mathbb{R}$. Set*

$$J_\phi(x, y) := \det \begin{pmatrix} \frac{\partial \phi_1}{\partial x} & \frac{\partial \phi_1}{\partial y} \\ \frac{\partial \phi_2}{\partial x} & \frac{\partial \phi_2}{\partial y} \end{pmatrix}.$$

*Then $J_\phi(x, y) \geq 0$ for every $x, y \in \mathbb{R}$.*

**Proof** Since

$$\phi'(z_0) = \lim_{h \to 0} \frac{\phi(z_0 + h) - \phi(z_0)}{h},$$

taking $h = t \in \mathbb{R}$ (resp. $h = it, t \in \mathbb{R}$), we obtain $\phi'(z_0) = \frac{\partial \phi_1}{\partial x} + i\frac{\partial \phi_2}{\partial x}$ (resp. $\phi'(z_0) = \frac{\partial \phi_2}{\partial y} - i\frac{\partial \phi_1}{\partial y}$) which yield the *Cauchy–Riemann equations*

$$\frac{\partial \phi_1}{\partial x} = \frac{\partial \phi_2}{\partial y} \quad \text{and} \quad \frac{\partial \phi_1}{\partial y} = -\frac{\partial \phi_2}{\partial y}.$$

We conclude that

$$J_\phi(x, y) = \left(\frac{\partial \phi_1}{\partial x}\right)^2 + \left(\frac{\partial \phi_1}{\partial y}\right)^2 \geq 0.$$

$\square$

**Theorem 2.20** *Let $D \subset \mathbb{C}$ be a connected, bounded, open set such that $\partial D$ is a $C^1$ closed curve. Let $\phi : \bar{D} \to \mathbb{C}$ be a holomorphic function and let $p \notin \phi(\partial D)$. Then*

$$d(\phi, D, p) = w(\phi, D, p) = \frac{1}{2\pi i} \int_{\partial D} \frac{\phi'(z)}{\phi(z) - p} \, dz.$$

**Proof** Suppose, first, that $\phi$ is constant, $\phi(z) = c$ for every $z \in D$. For every $p \in \mathbb{C} \setminus \phi(\partial D)$ we have $p \in \mathbb{C} \setminus \phi(\bar{D})$ and so $d(\phi, D, p) = 0$. Also, $w(\phi, D, p) = \frac{1}{2\pi i} \int_{\partial D} \frac{0 \, dz}{c - p} = 0$ and so

$$d(\phi, D, p) = w(\phi, D, p).$$

Now let $\phi$ be a nonconstant holomorphic function in $D$. If $p \notin \phi(\bar{D})$, then $d(\phi, D, p) = 0$ and $z \to \frac{\phi'(z)}{\phi(z) - p}$ is a holomorphic function on $\bar{D}$. Therefore, $\int_{\partial D} \frac{\phi'(z)}{\phi(z) - p} \, dz = 0$ and

$$d(\phi, D, p) = w(\phi, D, p).$$

Assume that $p \in \phi(D)$. Since $\phi(z)$ is nonconstant, $\phi^{-1}\{p\}$ must be finite (see Remark 2.16), say

$$\phi^{-1}\{p\} = \{z_1, \ldots, z_k\}.$$

By Lemma 2.18, for each $j = 1, \ldots, k$, there are $R_j > 0$, $\varphi_j : B(z_j, R_j) \to \mathbb{C}$ such that

$$\varphi_j \in H(B(z_j, R_j)), \quad B(z_j, R_j) \subset\subset D,$$
$$\varphi_j : B(z_i, R_j) \to \varphi_j(B(z_j, R_j)) \text{ is a bijection,}$$
$$\phi(z) - p = (z - z_j)^{m_j} g_j(z), \quad \text{for every } z \in B(z_j, R_j),$$

$$(\varphi_j)^{m_j} = (z - z_j)^{m_j} g_j(z) \text{ for every } z \in B(z_j, R_j),$$

for some $g_j \in H(B(z_j, R_j))$. If $R := \frac{1}{2} \min\{R_1, \ldots, R_k\}$, then

$$w(\phi, D, p) = \frac{1}{2\pi i} \int_{\partial D} \frac{\phi'(z)}{\phi(z) - p} \, dz = \sum_{j=1}^{k} \frac{1}{2\pi i} \int_{\partial B(z_j, R)} \frac{\phi'(z)}{\phi(z) - p} \, dz.$$

We now show that

$$\frac{1}{2\pi i} \int_{\partial B(z_j, R)} \frac{\phi'(z) \, dz}{\phi(z) - p} \, dz = m_j.$$

Indeed,

$$\phi(z) = p + (z - z_j)^{m_j} g_j(z)$$

and so

$$\phi'(z) = (z - z_j)^{m_j - 1} [m_j g_j(z) + (z - z_j) g_j'(z)].$$

This implies that

$$\int_{\partial B(z_j, R)} \frac{\phi'(z)}{\phi(z) - p} \, dz = \int_{\partial B(z_j, R)} \frac{m_j}{z - z_0} + \int_{\partial B(z_j, R)} \frac{g_j'(z)}{g_j(z)} \, dz = 2\pi i m_j;$$

therefore,

$$w(\phi, D, p) = \sum_{j=1}^{k} m_j.$$

It remains to show that $d(\phi, D, p) = \sum_{j=1}^{k} m_j$.

Let $C$ be the connected component of $\mathbb{C} \setminus \phi(\partial D)$ containing $p$. It is obvious that $C$ is an open set, hence there is $\delta_0 > 0$ such that $p + \delta \in C$ for every $|\delta| < \delta_0$ and we have

$$d(\phi, D, p) = d(\phi, D, p + \delta)$$

for every $|\delta| < \delta_0$. The equation

$$\phi(z) = p + \delta, \quad z \in B(z_j, R)$$

is equivalent to

$$p + (\varphi(z))^{m_j} = p + \delta, \quad z \in B(z_j, R),$$

which, in turn, is equivalent to

$$(\varphi(z))^{m_j} = |\delta| e^{2\pi i \theta},$$

where $\frac{\delta}{|\delta|} = e^{2\pi i \theta}$ and $\theta \in [0, 1[$. As $\varphi_j$ is injective on $B(z_j, R)$, the equation $(\varphi_j(z))^{m_j} = |\delta| e^{2\pi i \theta}$ has exactly $m_j$ distinct solutions in $B(z_j, R)$ for every $|\delta| < \delta_1$, for some $\delta_0 > \delta_1 > 0$. Let $z_j^1, \ldots, z_j^{m_j}$ be these solutions. We have

$$\phi'(z_j^l) = m_j(\varphi_j(z_j^l))^{m_j-1}\varphi_j'(z_j^l) \neq 0.$$

Using the definition of the degree for $C^1$ mappings by Lemma 2.19, we have

$$d(\phi, D, p + \delta) = \sum_{j=1}^{k} \sum_{l=1}^{m_j} \text{sgn}(J_\phi(z_j^l)) = \sum_{j=1}^{k} m_j$$

and we conclude that

$$d(\phi, D, p) = w(\phi, D, p).$$

$\square$

## 2.6   Exercises

**Exercise 2.1** Let $D = (-1, 1) \times (-1, 1)$ and let $\phi \in C(\bar{D})^2$ be defined by

$$\phi(x_1, x_2) = (\max\{|x_1|, |x_2|\}, 0), \quad \text{for } (x_1, x_2) \in \bar{D}.$$

Find $d(\phi, D, p)$ at $p = (0, 0)$.

*Solution 2.1.* First we note that $p = (0, 0) \notin \phi(\partial D)$ and so $d(\phi, D, p)$ is well defined. Let $\psi \in C^1(\bar{D})^2$ be defined by

$$\psi(x_1, x_2) = (1, 0), \quad \text{for } (x_1, x_2) \in \bar{D}.$$

We have $\phi|_{\partial D} = \psi|_{\partial D}$ and $p \notin \psi(\bar{D})$. By Theorem 2.1 we have $d(\psi, D, p) = 0$ and by Theorem 2.4 we deduce that $d(\phi, D, p) = 0$.

**Exercise 2.2** Let $D = (-1, 1) \times (-1, 1)$ and let $\phi \in C(\bar{D})^N$ be defined by

$$\phi(x_1, x_2) = (x_2 - x_1^3, x_2), \quad (x_1, x_2) \in \bar{D}.$$

Find $d(\phi, D, p)$ at $p = (0, 0)$.

*Solution 2.2.* We have $\phi(x_1, x_2) = (0, 0)$ if and only if $(x_1, x_2) = (0, 0)$. Therefore, $(0, 0) \notin \phi(\partial D)$ and so $d(\phi, D, p)$ is well defined. Also, $J_\phi(x_1, x_2) = -3x_1^2$ and so $(0, 0) \in \phi(Z_\phi)$. Setting $q = (-\frac{1}{2}, 0)$, then $q$ and $(0, 0)$ belong to the same connected component of $\mathbb{R}^2 \setminus \phi(\partial D)$. Moreover, $\phi(x_1, x_2) = q$ if and only if $(x_1, x_2) = (\frac{1}{2}^{\frac{1}{3}}, 0)$. Since $q \notin \phi(\partial D)$, it is easy to see that $d(\phi, D, q) = \sum_{x \in \psi^{-1}(q)} \text{sgn}(J_\phi(x)) = -1$; thus

$$d(\phi, D, p) = -1.$$

**Exercise 2.3** Let $a, b \in \mathbb{R}$ be such that $a < b$ and set $D = (a, b)$. Assume that $\phi \in C(\bar{D})$ and $p \notin \{\phi(a), \phi(b)\}$. Prove that $d(\phi, D, p) \in \{-1, 0, 1\}$.

*Solution 2.3.* We first note that $p \notin \phi(\partial D)$ and so $d(\phi, D, p)$ is well defined. Set

$$\psi(x) := \frac{\phi(b) - \phi(a)}{b - a}(x - a) + \phi(a).$$

If $\phi(b) = \phi(a)$, then $\psi$ is a constant and $d(\psi, D, p) = 0$. Assume that $\phi(b) \neq \phi(a)$. We obtain

$$d(\psi, D, p) = \begin{cases} \text{sgn}\dfrac{\phi(b) - \phi(a)}{b - a} & p \in [\phi(a), \phi(b)] \\ 0 & p \notin [\phi(a), \phi(b)] \end{cases}$$

and since $\phi|_{\partial D} = \psi|_{\partial D}$, by Theorem 2.4 we deduce that $d(\phi, D, p) = d(\psi, D, p)$ and so $d(\phi, D, p) \in \{-1, 0, 1\}$.

**Exercise 2.4 [Poincaré–Bohl]** Let $D \subset \mathbb{R}^N$ be an open, bounded set, $\phi, \psi \in C(\bar{D})^N$ and $p \notin [\phi(x), \psi(x)] := \{a\phi(x) + (1 - a)\psi(x) : a \in [0, 1]\}$ for every $x \in \partial D$. Prove that $d(\phi, D, p) = d(\psi, D, p)$.

*Solution 2.4.* It is obvious that $d(\phi, D, p)$ and $d(\psi, D, p)$ exist. Set

$$H(x, t) := t\phi(x) + (1 - t)\psi(x), \quad x \in \bar{D}, \ t \in [0, 1].$$

$H$ is a $C$ homotopy between $\psi$ and $\phi$ and $p \in H(\partial D, t)$ for every $t \in [0, 1]$. Therefore, Theorem 2.3 implies that $d(H(\cdot, t), D, p)$ does not depend on $t \in [0, 1]$ and so $d(\phi, D, p) = d(\psi, D, p)$.

# 3

# SOME APPLICATIONS OF THE DEGREE THEORY TO TOPOLOGY

## 3.1 The Brouwer Fixed Point Theorem

In this section we give two versions of the Brouwer Fixed Point Theorem.

**Definition 3.1** A set $E \subset \mathbb{R}^N$ is said to be a convex set *if for every* $x, y \in E$ *and for every* $\lambda \in (0, 1), \lambda x + (1 - \lambda)y \in E$.

**Definition 3.2** *Let* $E$ *be a convex set such that* $0 \in \text{int } E$. *We define the* gauge *of* $E$ *(or* Minkowski function*)* , $\rho_E$, *by*

$$\rho_E(x) := \inf \left\{ t > 0 \ : \ \frac{x}{t} \in E \right\}, \quad x \in \mathbb{R}^N.$$

**Remark 3.3** Since $0 \in \text{int } E$, there is $\eta > 0$ such that $\bar{B}(0, \eta) \subset E$. As $\frac{x}{t} \in E$ for every $t \geq \frac{|x|_2}{\eta}$, it follows that $\{t > 0 \ : \ \frac{x}{t} \in E\} \neq \emptyset$ and so $\rho_E(x)$ is well defined.

**Lemma 3.4** *Let* $E \subset \mathbb{R}^N$ *be a convex set containing* $0$ *in its interior. Let* $\rho_E$ *be the gauge of* $E$. *Then*

(1) $x \in E \implies \rho_E(x) \leq 1$;

(2) $\rho_E(\lambda x) = \lambda \rho_E(x)$ *for every* $\lambda > 0$ *and for every* $x \in \mathbb{R}^N$;

(3) *there exists* $m > 0$ *such that* $\rho_E(x) \leq m|x|_2$ *for all* $x \in \mathbb{R}^N$;

(4) $\rho_E(x + y) \leq \rho_E(x) + \rho_E(y)$ *for every* $x, y \in \mathbb{R}^N$;

(5) $\rho_E$ *is continuous in* $\mathbb{R}^N$;

(6) *if* $E$ *is bounded, then* $\rho_E(x) > 0$ *for every* $x \neq 0$ *and* $\frac{x}{\rho_E(x)} \in \bar{E}$.

**Proof**

(1) Clearly, if $x = \frac{x}{1} \in E$, then $\inf\{t > 0 \ : \ \frac{x}{t} \in E\} \leq 1$, i.e. $\rho_E(x) \leq 1$.

(2) Given $\lambda > 0$,

$$\rho_E(\lambda x) = \inf \left\{ t > 0 \ : \ \frac{\lambda x}{t} \in E \right\}$$

$$= \inf \left\{ \lambda s \ : \ \frac{\lambda x}{\lambda s} \in E, \ s > 0 \right\}$$

$$= \lambda \inf \left\{ s > 0 \ : \ \frac{x}{s} \in E \right\}$$

$$= \lambda \rho_E(x).$$

(3) Since $0 \in E$, there exists $\eta > 0$ such that $\bar{B}(0, \eta) \subset E$. For every $t \geq \frac{|x|_2}{\eta}$ we obtain that $\frac{x}{t} \in E$ and so $\rho_E(x) \leq \frac{|x|_2}{\eta}$.

(4) Let $x, y \in \mathbb{R}^N$. There exist two sequences $\{\alpha_n\}, \{\beta_n\} \subset (0, \infty)$ such that

$$\rho_E(x) = \lim_{n \to +\infty} \alpha_n, \quad \frac{x}{\alpha_n} \in E$$

and

$$\rho_E(y) = \lim_{n \to +\infty} \beta_n, \quad \frac{y}{\beta_n} \in E.$$

Since $E$ is a convex set, we deduce that

$$\frac{\alpha_n}{\alpha_n + \beta_n} \frac{x}{\alpha_n} + \frac{\beta_n}{\alpha_n + \beta_n} \frac{y}{\beta_n} \in E;$$

thus $\frac{x+y}{\alpha_n + \beta_n} \in E$ and so $\alpha_n + \beta_n \geq \rho_E(x + y)$. Passing to the limit we have

$$\rho_E(x) + \rho_E(y) \geq \rho_E(x + y).$$

(5) Let $x, h \in \mathbb{R}^N$. By (3) and (4) we have

$$\rho_E(x + h) \leq \rho_E(x) + \rho_E(h) \quad \Rightarrow \quad \rho_E(x + h) - \rho_E(x) \leq \rho_E(h) \leq \lambda |h|_2$$

and

$$\rho_E(x) \leq \rho_E(x + h) + \rho_E(-h) \quad \Rightarrow \quad -\rho_E(x + h) + \rho_E(x) \leq \lambda |h|_2.$$

Therefore,

$$|\rho_E(x + h) - \rho_E(x)| \leq \lambda |h|_2$$

and so $\rho_E$ is continuous in $\mathbb{R}^N$.

(6) Assume that $E$ is bounded and let $R > 0$ be such that $E \subset B(0, R)$. If $\frac{x}{t} \in E$, then $\frac{|x|_2}{R} < t$ and so

$$\rho_E(x) \geq \frac{|x|_2}{R} > 0 \quad \text{if } x \neq 0.$$

Also, $\frac{0}{t} \in E$ for every $t > 0$ implies that

$$\rho_E(0) = 0.$$

Finally, considering a sequence $\{\alpha_n\} \subset (0, \infty)$ such that $\rho_E(x) = \lim \alpha_n$ and $\frac{x}{\alpha_n} \in E$, taking the limit as $n$ goes to infinity, we conclude that

$$\frac{x}{\rho_E(x)} \in \bar{E}. \qquad \square$$

**Proposition 3.5** *Let $E \subset \mathbb{R}^N$ be a compact, convex set such that $0 \in \text{int } E$. There exists a homeomorphism $\alpha : \mathbb{R}^N \to \mathbb{R}^N$ such that*

$$\alpha(E) = \bar{B},$$

*where $B$ is the unit ball of $\mathbb{R}^N$ with respect to the norm $|\cdot|_2$.*

**Proof**  Define $\alpha, \beta : \mathbb{R}^N \to \mathbb{R}^N$ by

$$\alpha(x) := \begin{cases} \frac{\rho_E(x)}{|x|_2} x & x \neq 0 \\ 0 & x = 0, \end{cases} \qquad \beta(y) := \begin{cases} \frac{|y|_2}{\rho_E(y)} y & y \neq 0 \\ 0 & y = 0. \end{cases}$$

For every $y \neq 0$ we have $\alpha \circ \beta(y) = \alpha(\frac{|y|_2}{\rho_E(y)} y) = \frac{\rho_E(y)}{|y|_2} \frac{|y|_2}{\rho_E(y)} y = y$ and $\alpha \circ \beta(0) = 0$. Thus $\alpha \circ \beta(y) = y$ for every $y \in \mathbb{R}^N$ and, similarly, $\beta \circ \alpha(x) = x$ for every $x \in \mathbb{R}^N$. Therefore, $\alpha : \mathbb{R}^N \to \mathbb{R}^N$ is a bijection. From Lemma 3.4 (5) it follows that $\alpha$ is continuous at every $x \neq 0$. Moreover,

$$|\alpha(x) - \alpha(0)|_2 = \left| \frac{\rho_E(x)}{|x|_2} x \right|_2 = \rho_E(x);$$

therefore $\lim_{x \to 0} \alpha(x) = 0$ and $\alpha$ is continuous at 0. Hence, $\alpha$ is continuous on $\mathbb{R}^N$. Similarly, $\beta : \mathbb{R}^N \to \mathbb{R}^N$ is continuous.

Next, we prove that $\alpha(E) = \bar{B}$. By Lemma 3.4 (1), $x \in E$ implies that $|\alpha(x)|_2 = \rho_E(x) \leq 1$, therefore $\alpha(E) \subset \bar{B}(0,1)$. Conversely, let $y \in \bar{B}(0,1)$. Then $y = \alpha(\beta(y))$ and $\rho_E(\beta(y)) = |y|_2 \leq 1$ and by Lemma 3.4 (6) we have $\frac{\beta(y)}{\rho_E(\beta(y))} \in E$. Since $E$ is a convex set containing 0, we obtain

$$(1 - \rho_E \circ \beta(y)) 0 + \rho_E \circ \beta(y) \frac{\beta(y)}{\rho_E \circ \beta(y)} \in E,$$

i.e. $\beta(y) \in E$. Thus $y \in \alpha(E)$ and so $\bar{B}(0,1) \subset \alpha(E)$.  □

**Remark 3.6**  Using the gauge function it can be shown that if $\Omega \subset \mathbb{R}^N$ is a smooth domain (e.g. strongly Lipschitz) and if $\mathcal{L}^N(\Omega) = \mathcal{L}^N(B(0,1))$, then there exists a Lipschitz map $v : \Omega \to B$ such that

(i) $v(\Omega) = B$;

(ii) $\det \nabla v = 1$ $\mathcal{L}^N$ a.e. $x \in \Omega$;

(iii) $v \in C^1(U_i)$ for some finite partition $\{U_i\}_{i=1,\dots,M}$ of $\Omega$ into smooth domains (e.g. strongly Lipschitz).

For a proof of this result, we refer the reader to Fonseca and Parry (1987).

The next two results are known as Brouwer's Fixed Point Theorem, the first one being the most commonly stated, while the latter has wider applications.

**Theorem 3.7** [**First Version of the Brouwer Fixed Point Theorem**] *Let* $D \subset \mathbb{R}^N$ *be an open, bounded set such that* $\bar{D}$ *is homeomorphic to the closed unit ball* $\bar{B}$. *Let* $\phi \in C(\bar{D})^N$ *be such that* $\phi(\bar{D}) \subset \bar{D}$. *Then* $\phi$ *has a fixed point in* $\bar{D}$.

**Proof** Let $\alpha : \bar{D} \to \bar{B}$ be a homeomorphism. Setting $\psi := \alpha \circ \phi \circ \alpha^{-1}$, we show that $\psi : \bar{B} \to \bar{B}$ has a fixed point. Clearly, $\psi$ is continuous and either there exists $x \in \partial B$ such that $\psi(x) = x$, in which case $\psi$ admits a fixed point, or $\psi(x) \neq x$ for every $x \in \partial\bar{B}$. Define

$$H(x, t) := x - t\psi(x) \text{ for } x \in \bar{B}, t \in [0, 1].$$

Then $0 \notin H(\partial B, t)$ for every $t \in [1, 0]$ and so, by Theorem 2.3 (2), $d(H(\cdot, t), B, 0)$ does not depend on $t$. This yields

$$
\begin{aligned}
d(I - \psi, B, 0) &= d(H(\cdot, 1), B, 0) \\
&= d(H(\cdot, 0), B, 0) \\
&= d(I, B, 0) = 1.
\end{aligned}
$$

By Theorem 2.7 the equation $(I - \psi)(x) = 0$ admits a solution in $B$, i.e. $\psi$ has a fixed point $\bar{x} \in \bar{B}$. Setting $y = \alpha^{-1}(x)$, we have $y \in \bar{D}$ and

$$\phi(y) = y.$$

$\square$

**Corollary 3.8** [**Second Version of Brouwer Fixed Point Theorem**] *Let* $K \subset \mathbb{R}^N$ *be a compact, convex set such that the interior of* $K$ *is not empty. Let* $\phi \in C(K)^N$ *be such that* $\phi(K) \subset K$. *Then* $\phi$ *admits a fixed point.*

**Proof** Let $x_0 \in \text{int } K$ and define $T : \mathbb{R}^N \to \mathbb{R}^N$ by $T(x) = x - x_0$. Then $T(K)$ is a compact, convex set such that $0 \in \text{int } T(K)$. By Proposition 3.5 there exists a homeomorphism $\alpha : T(K) \to \bar{B}$ and we define

$$\psi := T \circ \phi \circ T^{-1}.$$

Then $\psi : T(K) \to T(K)$ is continuous and by Theorem 3.7 $\psi$ has a fixed point $y \in T(K)$. Let $x \in K$ be such that $y = T(x)$. It is clear that $x$ is a fixed point of $\phi$. $\square$

We give some applications of the Brouwer Fixed Point Theorem.

**Proposition 3.9** *Assume that* $N$ *is odd,* $0 \in D$, *and* $\phi \in C(\bar{D})^N$ *is such that* $0 \notin \phi(\partial D)$. *Then there exist* $y \in \partial D$, $\lambda \neq 0$, *such that* $\phi(y) = \lambda y$.

**Proof** If for every $y \in \partial D$, $\lambda \in \mathbb{R}$, we have $\phi(y) \neq \lambda y$, then set

$$H(x, t) := tx + (1 - t)\phi(x), \quad x \in \bar{D}, \quad t \in [0, 1]$$

and

$$K(x,t) := -tx + (1-t)\phi(x), \quad x \in \bar{D}, \ t \in [0,1].$$

We have $0 \notin H(\partial D, t)$ for every $t \in [0,1]$ and $0 \notin K(\partial D, t)$ for every $t \in [0,1]$. Therefore, by Theorem 2.3, $d(H(0,t), D, 0)$ and $d(K(\cdot, t), D, 0)$ are independent of $t$ and so, as $H(\cdot, 0) = \varphi(\cdot) = K(\cdot, 0)$,

$$1 = d(I, D, 0) = d(-I, D, 0) = (-1)^N = -1,$$

which yields a contradiction. We conclude that there exist $y \in \partial D$ and $\lambda \in \mathbb{R}$ such that $\phi(y) = \lambda y$ and, since $0 \in \phi(\partial D)$, we must have $\lambda \neq 0$. □

**Theorem 3.10 [Perron–Frobenius]** *Let $A = (a_{ij})$ be an $N \times N$ matrix such that $a_{ij} \geq 0$ for all $i, j$. Then there exist $\lambda \geq 0$, $x \neq 0$, such that $x_i \geq 0$ for every $i$ and $Ax = \lambda x$.*

**Proof** See Exercice 3.1. □

**Remark 3.11** In Proposition 3.9 the condition '$N$ is odd' is essential. Indeed, let $N = 2$ and define $\phi : \bar{B} \to \mathbb{R}^2$ in polar co-ordinates by $\phi(r, \theta) = (r, \theta + r)$. The equation

$$(\cos(\theta + 1), \sin(\theta + 1)) = \lambda(\cos\theta, \sin\theta)$$

has no solution.

The next application of Proposition 3.9 provides periodic solutions for Lipschitz flows. We will need some standard results on ordinary differential equations, which we will state without proving them.

**Theorem 3.12** *Let $f : B(0, r) \times \mathbb{R} \to \mathbb{R}^N$ be a locally Lipschitz function. Consider the initial value problem*

$$\begin{cases} \dot{x} = f(x, t) \\ x(t_0) = x_0, \end{cases}$$

*where $t_0 \in \mathbb{R}$, $x_0 \in B(0, r)$. Then there exists a function*

$$x : U \to \mathbb{R}^N, \quad (t; t_0, x_0) \mapsto x(t, t_0, x_0),$$

*where $U := \{(t, t_0, x_0) : (t_0, x_0) \in \mathbb{R} \times B(0, r), \ t \in I(t_0, x_0)\}$, such that*

(i) *for each $(t_0, x_0) \in \mathbb{R} \times B(0, r)$, $x(t; t_0, x_0)$ is continuously differentiable with respect to $t$, and it satisfies (3.12) in the open interval $I(t_0, x_0)$;*

(ii) *if $y$ is a solution of (3.12) defined in $(\alpha, \beta)$, then $(\alpha, \beta) \subset I(t_0, x_0)$ and $y(t) = x(t, t_0, x_0)$ for every $t \in (\alpha, \beta)$;*

(iii) *if $I(t_0, x_0) = (\tau, \sigma)$, then either $\sigma = r$ or $\lim_{t \to \sigma} |x(t, t_0, x_0)| = r$;*

(iv) *for every $(t_0, x_0) \in \mathbb{R} \times B(0, r)$, there exists a $\delta > 0$ such that $|s_0 - t_0| + |y_0 - x_0| < \delta$ implies that $I(t_0, x_0) \subset I(s_0, y_0)$.*

**Definition 3.13** *Assume that there exist $T > 0$, $\phi : \mathbb{R} \to \mathbb{R}$, such that*

$$\dot{\phi} = f(\phi, t), \quad \phi(t + T) = \lambda \phi(t),$$

*for every $t \in \mathbb{R}$ and for some constant $\lambda \neq 0$. Then $\phi$ is called a* Floquet solution *of (3.12). Moreover, if $\lambda = 1$, $\phi$ is said to be* periodic of the first kind *and if $\lambda \neq 1, 0$, $\phi$ is said to be* periodic of the second kind.

The task ahead will be to prove that if $f(x, \cdot)$ is periodic for every $x \in \mathbb{R}^N$ and if $|x_0|_2$ is small enough, then there exists a Floquet solution for the initial value problem

$$\begin{cases} \dot{\phi} = f(\phi, t) \\ \phi(0) = x_0. \end{cases}$$

**Theorem 3.14** *Let $f : \mathbb{R}^N \times \mathbb{R} \to \mathbb{R}^N$ be a locally Lipschitz function such that $f(\cdot, t)$ is positively homogeneous of degree one (i.e. $f(\gamma x, t) = \gamma f(x, t)$ for every $\gamma > 0$ and for every $(x, t) \in \mathbb{R}^N \times \mathbb{R})$. Assume that $f(x, \cdot)$ is periodic, of period $T > 0$, for every $x \in \mathbb{R}^N$. Then there exists a $\delta > 0$ such that for every $0 < \alpha \leq \delta$ there exists $y_\alpha \in \bar{B}(0, \alpha)$ such that*

$$\begin{cases} \dot{x} = f(x, t) \\ x(0) = y_\alpha, \end{cases}$$

*admits a Floquet solution. Futhermore, if $N$ is odd, then $y_\alpha$ can be taken on $\partial B(0, \alpha)$.*

**Proof** First we suppose that $N$ is odd. As $f$ is continuous and positively homogeneous of degree one, we must have $f(0, t) = 0$ for every $t \in \mathbb{R}$. Therefore, by Theorem 3.12 (ii), $x(t) \equiv 0$ for all $t \in \mathbb{R}$ is the unique solution of the equation

$$\begin{cases} \dot{x} = f(x, t) \\ x(0) = 0. \end{cases}$$

Let us denote by $x(t, 0, c)$ the solution of the equation

$$\begin{cases} \dot{x} = f(x, t) \\ x(0) = c. \end{cases} \tag{E}$$

By Theorem 3.12 (iv) there exists $\delta > 0$ such that $|c|_2 \leq \delta$ implies that the equation (E) admits a unique solution defined in $\mathbb{R}$. Fix $0 < \alpha \leq \delta$ and define $F : \bar{B}(0, \alpha) \to \mathbb{R}^N$ by

$$F(c) := x(T, 0, c).$$

By Theorem 3.12 $F$ is continuous and by the uniqueness of the solutions of (E), $F(c) = 0$ if and only if $c = 0$. Therefore, $0 \notin F(\partial B(0, \alpha))$ and, by Proposition 3.9, there are $c_\alpha \in \partial B(0, \alpha)$, $\lambda > 0$, such that $T(c_\alpha) = \lambda c_\alpha$. For $t \in \mathbb{R}$ set

$$\phi(t) := \lambda x(t, 0, c_\alpha)$$

and
$$\psi(t) := x(t + T, 0, c_\alpha).$$
It suffices to prove that $\phi(t) = \psi(t)$ for every $t \in \mathbb{R}$. Indeed,
$$
\begin{aligned}
\dot{\phi}(t) &= \lambda \dot{x}(t, 0, c_\alpha) \\
&= \lambda f(x(t, 0, c_\alpha), t) \\
&= f(\lambda x(t, 0, c_\alpha), t) \\
&= f(\phi(t), t)
\end{aligned}
$$
and $\phi(0) = \lambda x(0, 0, c_\alpha) = \lambda c_\alpha$. Also,
$$
\begin{aligned}
\dot{\psi}(t) &= \dot{x}(t + T, 0, c_\alpha) \\
&= f(x(t + T, 0, c_\alpha), t) \\
&= f(x(t + T, 0, c_\alpha), t + T) \\
&= f(\psi(t), t + T) \\
&= f(\psi(t), t)
\end{aligned}
$$
and
$$\psi(0) = x(T, 0, c_\alpha) = Tc_\alpha = \lambda c_\alpha.$$
By the uniqueness of the solution of (E), we conclude that $\phi(t) = \psi(t)$ for every $t \in \mathbb{R}$.

Finally, if $N$ is even we define $g : \mathbb{R}^{N+1} \to \mathbb{R}^{N+1}$ by
$$g(x, x_{N+1}, t) := (f(x, t), x_{N+1}).$$
Since $N + 1$ is odd, there exists $(y, y_{N+1}) \in \mathbb{R}^{N+1}$, $|(y, y_{N+1})|_2 = \alpha$, for which
$$
\begin{cases}
\begin{pmatrix} \dot{x} \\ \dot{x}_{N+1} \end{pmatrix} = g(x, x_{N+1}, t) \\[2mm]
\begin{pmatrix} x(0) \\ x_{N+1}(0) \end{pmatrix} = \begin{pmatrix} y \\ y_{N+1} \end{pmatrix}
\end{cases}
$$
admits a Floquet solution. We conclude that (E) admits a Floquet solution for some initial value $y$, $|y|_2 \leq \alpha$. $\qquad\square$

## 3.2 Odd mappings

The aim of this section is to show that the degree of a continuous, odd function $\phi : \Omega \to \mathbb{R}^N$ is an odd number (see Theorem 3.23). As a consequence of this result, if $\phi : \bar{B}(0, 1) \subset \mathbb{R}^N \to \mathbb{R}^N$ is an odd function then the equation $\phi(x) = 0$ admits a solution in $\bar{B}(0, 1)$. Indeed, since $d(\phi, B(0, 1), 0)$ is odd, we obtain
$$d(\phi, B(0, 1), 0) \neq 0$$
and using Theorem 2.1 we conclude that the equation $\phi(x) = 0$ admits a solution.

**Definition 3.15** *The set $D \subset \mathbb{R}^N$ is said to be* symmetric *if for every $x \in D$ we have $-x \in D$. The mapping $\phi : D \to \mathbb{R}^M$ is said to be an* odd *mapping if $\phi(x) = -\phi(-x)$ for every $x \in D$.*

Before stating the main result of this section, the Odd Mapping Theorem (Theorem 3.23), following the scheme of Schwartz (1969) we first make some remarks on the Tietze Extension Theorem (see Theorem 1.15) and we prove some lemmas which will be used in the sequel.

**Remark 3.16** We may deduce from Tietze's Extension Theorem (Theorem 1.15) that if $K, L \subset \mathbb{R}^N$ are two compact sets such that $K \subset L$ and if $f : K \to \mathbb{R}^M$ is a continuous function, then there is a continuous extension $g : L \to \mathbb{R}^N$ of $f$ such that

$$|f|_K = |g|_L = \sup\{|g_i|_L : i = 1, \ldots, M\}.$$

This is Exercise 3.2.

**Proposition 3.17** *Let $M > N$ be two integers, let $K \subset \mathbb{R}^N$ be a compact set, and let $\phi \in C^1(K)^M$. Then $\phi(K)$ has measure zero in $\mathbb{R}^M$.*

**Proof** We identify $K$ with $K' \subset \mathbb{R}^M$ and $\phi : K \to \mathbb{R}^M$ with $\psi : K' \to \mathbb{R}^M$, where

$$K' := \{(x, 0, \ldots, 0) \in K \times \mathbb{R}^{M-N}\}$$

and

$$\psi(x, 0, \ldots, 0) := \phi(x).$$

Since $J_\psi(y) = 0$ for every $y \in K'$, by Sard's Lemma 1.4 we have

$$\mathcal{L}^M(\phi(K)) = \mathcal{L}^M(\psi(K')) = 0.$$

$\square$

**Theorem 3.18** *Let $K \subset L \subset \mathbb{R}^N$ be two compact, nonempty sets, let $M > N$ be two integer numbers, and let $\phi \in C(K)^M$ be a function such that $\phi$ is nowhere zero. Then $\phi$ can be extended to a mapping $\chi \in C(L)^M$ which is nowhere zero.*

**Proof** Let $c := \min\{|\phi(x)| : x \in K\} > 0$ and choose $0 < \epsilon < \frac{c}{2}$.
*Claim 1.* We claim that there exists $\psi \in C^1(L)^M$ such that $\psi$ is nowhere zero and $|\psi(x) - \phi(x)| < \epsilon$ for every $x \in K$.

Indeed, by Tietze's Extension Theorem (see Theorem 1.15), there exists $\phi_1 \in C(L)^M$ such that

$$\phi_1|_K = \phi.$$

Let $\phi_2 \in C^1(L)^M$ be such that

$$|\phi_2(x) - \phi_1(x)| < \frac{\epsilon}{2}$$

for every $x \in L$. By Proposition 3.17,

$$\mathcal{L}^M(\phi_2(L)) = 0$$

and so there exists $p \notin \phi_2(L)$ such that $|p| < \frac{\epsilon}{2}$. Let

$$\psi(x) := \phi_2(x) - p.$$

Clearly, $\psi$ is nowhere zero, $\psi \in C^1(L)^M$, and, for every $x \in K$,

$$|\psi(x) - \phi(x)| \leq |\psi(x) - \phi_2(x)| + |\phi_2(x) - \phi(x)| < \frac{\epsilon}{2} + \frac{\epsilon}{2} = \epsilon.$$

*Claim 2.* There exists $\Phi \in C(L)^M$ such that $|\Phi(x) - \phi(x)| < \epsilon$ for every $x \in K$ and $|\Phi(x)| \geq \frac{c}{2}$ for every $x \in L$.
   Indeed, let

$$\eta(t) := \begin{cases} 1 & t \geq \frac{c}{2} \\ \frac{2t}{c} & t < \frac{c}{2}. \end{cases}$$

We obtain that $\eta \in C(\mathbb{R})$. Setting $\Phi(x) := \frac{\psi(x)}{\eta(|\psi(x)|)}$, clearly $\Phi \in C(L)^M$ and if $|\psi(x)| < \frac{c}{2}$, then $\Phi(x) = \frac{\psi(x)}{\frac{2|\psi(x)|}{c}}$ and so $|\Phi(x)| = \frac{c}{2}$. If $|\psi(x)| \geq \frac{c}{2}$, then $\Phi(x) = \psi(x)$ and so $|\Phi(x)| = |\psi(x)| \geq \frac{c}{2}$. Therefore, $|\Phi(x)| \geq \frac{c}{2}$ for every $x \in L$. Also, for every $x \in K$, we have

$$|\psi(x)| \geq |\phi(x)| - |\phi(x) - \psi(x)| \geq c - \frac{c}{2} = \frac{c}{2},$$

which implies that $\Phi(x) = \psi(x)$ and so

$$|\Phi(x) - \phi(x)| < \epsilon.$$

*Claim 3.* There exists $\chi \in C(L)^M$ such that $\chi|_K = \phi$ and $\chi$ is nowhere zero.
   Indeed, applying Remark 3.16 to $\Phi - \phi$ on $K$, we obtain $\alpha \in C(L)^M$ so that $|\alpha(x)| \leq \epsilon$ for every $x \in L$ and $\alpha|_K = \Phi - \phi$. Let

$$\chi := \Phi - \alpha.$$

We have

$$\chi|_K = \Phi - \alpha|_K = \Phi|_K - \phi|_K + \phi|_K = \phi$$

and $\chi \in C(L)^M$. Also, for every $x \in L$,

$$|\chi(x)| = |\Phi(x) - \alpha(x)| \geq |\Phi(x)| - |\alpha(x)| > \frac{c}{2} - \epsilon > 0.$$

$\square$

**Remark 3.19** The conclusion of Theorem 3.18 may not hold without the assumption $M > N$. Indeed, let $K = \partial B(0,1)$ and let $L = \bar{B}(0,1)$. Denote by $I$

the identity function on $\mathbb{R}^N$ and let $\phi \in C(K)^M$ be such that $\phi|_K = I|_K$. Then if $\chi$ is any extension of $\phi$ to $L$ we have

$$d(\chi, B(0,1), 0) = d(I, B(0,1), 0) = 1$$

and by Theorem 2.1 the equation $\chi(x) = 0$ must admit a solution.

**Lemma 3.20** *Let $M > N$ be two integer numbers and let $D \subset \mathbb{R}^N$ be a bounded, symmetric, open set such that $0 \notin \bar{D}$. Assume that $\phi : \partial D \to \mathbb{R}^M$ is a continuous, odd mapping and nowhere zero. Then there exists an odd and nowhere zero mapping $\psi : \bar{D} \to \mathbb{R}^M$ which extends $\phi$.*

We observe that since $D$ is a symmetric set, $\partial D$ and $\bar{D}$ are also symmetric sets.

**Proof** We proceed with the proof by induction on $N$. Assume first that $N = 1$. Since $0 \notin \bar{D} \subset \mathbb{R}$ and $D$ is a compact set, there are two real numbers $\epsilon$ and $A$ such that $0 < \epsilon < A$ and $D \subset [-A, -\epsilon] \cup [\epsilon, A]$. Let $\phi_1 = \phi|_{\partial D \cap [\epsilon, A]}$. Then $\phi$ is a continuous function which is nowhere zero and, by Theorem 3.18, there exists $\phi_2 : [\epsilon, A] \to \mathbb{R}^M$, nowhere zero and a continuous extension of $\phi_1$. Let

$$\psi(x) := \begin{cases} \phi_2(x) & x \in \bar{D} \cap [\epsilon, A] \\ -\phi_2(-x) & x \in \bar{D} \cap [-A, -\epsilon]. \end{cases}$$

The function $\psi : \bar{D} \to \mathbb{R}^M$ is continuous, odd and nowhere zero; furthermore,

$$\psi|_{\partial D} = \phi.$$

Now suppose that the $N > 1$ and that the lemma holds for $N-1$. For convenience, we identify $\mathbb{R}^{N-1}$ with $\mathbb{R}^N \cap \{x \in \mathbb{R}^N : x_N = 0\}$. Let

$$D_+ := \{x \in D : x_N > 0\}, \quad D_- := \{x \in D : x_N < 0\}.$$

We have

$$\bar{D} = D_+ \cup D_- \cup \partial D \cup (\bar{D} \cap \mathbb{R}^{N-1}).$$

Let

$$\phi_1 = \phi|_{\partial D \cap \mathbb{R}^{N-1}}.$$

By the induction hypothesis, since $\phi_1$ is an odd, nowhere zero, continuous mapping there exists $\phi_2 : \bar{D} \cap \mathbb{R}^{N-1} \to \mathbb{R}^M$, a continuous mapping which is odd, nowhere zero, and extends $\phi_1$. Let

$$\phi_3(x) := \begin{cases} \phi_2(x) & x \in \bar{D} \cap \mathbb{R}^{N-1} \\ \phi(x) & x \in \partial D. \end{cases}$$

We show that $\phi_3$ is continuous, nowhere zero, and odd. Note that $\phi_3$ is well defined. Indeed, if $x \in (\bar{D} \cap \mathbb{R}^{N-1}) \cap \partial D = \partial D \cap \mathbb{R}^{N-1}$, we have

$$\phi_2(x) = \phi_1(x) = \phi(x).$$

Also, since $\phi_2(x) \neq 0$ for every $x \in \bar{D} \cap \mathbb{R}^{N-1}$ and $\phi(x) \neq 0$ for every $x \in \partial D$, we obtain that

$$\phi_3(x) \neq 0$$

for every $x \in (\bar{D} \cap \mathbb{R}^{N-1}) \cup \partial D$. Moreover, if $x \in \bar{D} \cap \mathbb{R}^{N-1}$, then $-x \in \bar{D} \cap \mathbb{R}^{N-1}$ and

$$\phi_3(x) = \phi_2(x), \quad \phi_3(-x) = \phi_2(-x) = -\phi_2(x) \quad \text{and} \quad \phi_3(x) = -\phi_3(-x).$$

It is easy to see that for every $x \in \partial D$, $\phi_3(x) = \phi(x)$, $\phi_3(-x) = \phi(-x) = \phi(x)$, and $\phi_3(x) = -\phi_3(-x)$. Clearly, $\phi_3$ is continuous and since $\hat{D} = \bar{D} \setminus D_-$, $\hat{D} = (\bar{D} \cap \mathbb{R}^{N-1}) \cup D_+$ is a compact set, by Theorem 3.18 there exists $\phi_4 : \hat{D} \to \mathbb{R}^M$ that extends $\phi_3$, nowhere zero and continuous. Finally, define

$$\psi(x) := \begin{cases} \phi_4(x) & x \in \hat{D} \\ -\phi_4(-x) & x \in \bar{D} \setminus \hat{D} = D_-. \end{cases}$$

Then $\psi$ is nowhere zero and for every $x \in \partial D$ we have $x \in \hat{D}$, and so

$$\psi(x) = \phi_4(x) = \phi(x).$$

In order to prove that $\psi$ is an odd function, we observe that if $x \in \partial D$, then $-x \in \partial D$, $\psi(-x) = \phi(-x) = -\phi(x)$, and $\psi(x) = \phi(x)$, and we obtain

$$\psi(-x) = -\psi(x).$$

If $x \in D_+$, then $-x \in D_-$, $\psi(x) = \phi_4(x)$, and $\psi(-x) = -\phi_4(x)$, and so

$$\psi(x) = -\psi(-x).$$

If $x \in D_-$, then $-x \in D_+$ and, by an argument similar to the one above, we conclude that

$$\psi(x) = -\psi(-x).$$

Now we prove the continuity of $\psi$ on $\bar{D}$. If $x_0 \in D_+ \cup D_-$, it is clear that $\psi$ is continuous at $x_0$. It remains to study the case where $x_0 \in \partial D \cup (\bar{D} \cap \mathbb{R}^{N-1})$. Assume that

$$x_0 = \lim_{r \to +\infty} x_r,$$

where $\{x_r\} \subset \bar{D}$. Either there exists a subsequence $\{x_{r_k}\}$ such that $x_{r_k} \in \hat{D}$ or there exists a subsequence $\{\tilde{x}_{r_k}\}$ such that $\tilde{x}_{r_k} \notin \hat{D}$, i.e. $-\tilde{x}_{r_k} \in D_+$. This implies that either

$$\lim_{k \to +\infty} \psi(x_{r_k}) = \lim_{k \to +\infty} \phi_4(x_{r_k}) = \phi_4(x_0) = \psi(x_0)$$

or

$$\lim_{k \to +\infty} \psi(\tilde{x}_{r_k}) = -\lim_{k \to +\infty} -\phi_4(-\tilde{x}_{r_k}) = -\lim_{k \to +\infty} \phi_4(-x_0) = -\psi(-x_0) = \psi(x_0).$$

Hence $\psi$ is continuous.                                                    □

**Lemma 3.21** *Let $D \subset \mathbb{R}^N$ be a bounded, open, symmetric set such that $0 \notin \bar{D}$. Let $\phi : \partial D \to \mathbb{R}^N$ be a continuous, odd, nowhere zero mapping. Then $\phi$ can be an extended to a function $\psi : \bar{D} \to \mathbb{R}^N$ which is continuous, odd, and such that $\psi(x) \neq 0$ for every $x \in \bar{D} \cap \mathbb{R}^{N-1}$.*

**Proof**  We recall that we identify $\mathbb{R}^{N-1}$ with $\mathbb{R}^N \cap \{(x_1, \ldots, x_N) \in \mathbb{R}^N : x_N = 0\}$. We denote by $\mathbb{R}^N_+$ the set $\{x \in \mathbb{R}^N : x_N > 0\}$ and by $\mathbb{R}^N_-$ the set $\{x \in \mathbb{R}^N : x_N < 0\}$. Let

$$\phi_1 := \phi|_{\partial D \cap \mathbb{R}^{N-1}} = \phi|_{\partial(D \cap \mathbb{R}^{N-1})}.$$

As $\phi_1$ is a continuous, nowhere zero, odd mapping by Lemma 3.20, $\phi_1$ admits an extension $\phi_2 : \bar{D} \cap \mathbb{R}^{N-1} \to \mathbb{R}^N$ which is continuous, odd, and nowhere zero. We define $\phi_3 : (\bar{D} \cap \mathbb{R}^{N-1}) \cup \partial D \to \mathbb{R}^N$ by

$$\phi_3(x) := \begin{cases} \phi_2(x) & x \in \bar{D} \cap \mathbb{R}^{N-1} \\ \phi(x) & x \in \partial D. \end{cases}$$

The function $\phi_3$ is odd and nowhere zero. Note that if $x \in \partial D \cap (\bar{D} \cap \mathbb{R}^{N-1})$, then $x \in \partial D \cap \mathbb{R}^{N-1}$ and so $\phi_3$ is well defined because

$$\phi_2(x) = \phi_1(x) = \phi(x).$$

Since $\phi_3$ is continuous on the compact sets $\partial D$ and $\bar{D} \cap \mathbb{R}^{N-1}$, we deduce that $\phi_3$ is continuous on $\partial D \cup (\bar{D} \cap \mathbb{R}^{N-1})$. Also, since the compact set $\partial D \cup (\bar{D} \cap \mathbb{R}^{N-1})$ is a subset of the compact set $\partial D \cup (\bar{D} \cap \mathbb{R}^{N-1}) \cup (\bar{D} \cap \mathbb{R}^N_+) = \hat{D}$, by Remark 3.16 there exists a continuous extension of $\phi_3$, $\phi_4 : \hat{D} \to \mathbb{R}^N$ . We set

$$\psi(x) := \begin{cases} \phi_4(x) & x \in \hat{D} \\ -\phi_4(-x) & x \in \bar{D} \setminus \hat{D}. \end{cases}$$

The same argument used in the proof of Lemma 3.20 allows us to deduce that $\psi$ is continuous and odd, and as

$$\psi|_{\bar{D} \cap \mathbb{R}^{N-1}} = \phi_4|_{\bar{D} \cap \mathbb{R}^{N-1}} = \phi_3|_{\bar{D} \cap \mathbb{R}^{N-1}} = \phi_2,$$

we conclude that $\psi$ is nowhere zero on $\bar{D} \cap \mathbb{R}^{N-1}$.                    □

**Remark 3.22**  In Lemma 3.21 the extension $\psi : \bar{D} \to \mathbb{R}^N$ is not necessarily nowhere zero in $\bar{D}$. Indeed, let

$$D_1 := (-3, -1) \times (-1, 1)^{N-1}, \ D_2 := (1, 3) \times (-1, 1)^{N-1}$$

and
$$D := D_1 \cup D_2.$$

We observe that $D$ is an open, bounded, symmetric set of $\mathbb{R}^N$ such that $0 \notin \bar{D}$. Define $f : \bar{D} \to \mathbb{R}^N$ by

$$f(x) = \begin{cases} x + (2, 0, \ldots, 0) & x \in D_1 \\ x + (-2, 0, \ldots, 0) & x \in D_2. \end{cases}$$

and let $\phi := f|_{\partial D} = f|_{\partial D_1 \cup \partial D_2}$. Then any continuous extension of $\phi$ is zero somewhere in $D$.

For the sake of illustration, we give the proof in the case where $N = 2$. It is obvious that $\phi$ is a continuous, nowhere zero, odd mapping. Let $\psi : \bar{D} \to \mathbb{R}^N$ be any continuous extension of $\phi$. For $i = 1, 2$ we obtain that

$$d(\psi|_{D_i}, D_i, 0) = d(f|_{D_i}, D_i, 0) = 1.$$

and so the equation $\psi(x) = 0$ admits at least two distinct solutions in $D$.

Now we state and prove the main theorem of this section.

**Theorem 3.23 [The Odd Mapping Theorem]** *Let $D \subset \mathbb{R}^N$ be a bounded, open, symmetric set such that $0 \notin D$. Let $\phi \in C(\bar{D})^N$ be such that $0 \notin \phi(\partial D)$ and assume that*

$$\frac{\phi(x)}{|\phi(x)|} \neq \frac{\phi(-x)}{|\phi(-x)|}$$

*for every $x \in \partial D$. Then $d(\phi, D, 0)$ is an odd number.*

**Remark 3.24** If $\phi \in C(\bar{D})^N$ is an odd mapping such that $0 \notin \phi(\partial D)$, then

$$\frac{\phi(x)}{|\phi(x)|} \neq \frac{\phi(-x)}{|\phi(-x)|},$$

for every $x \in \partial D$ and the Odd Mapping Theorem applies.

**Proof of Theorem 3.23** We divide the proof into four steps.
*First step.* We show that we can assume, without loss of generality, that $\phi$ is an odd mapping. Indeed, let $\psi : \bar{D} \to \mathbb{R}^N$ be defined by

$$\psi(x) := \phi(x) - \phi(-x).$$

Then $\psi$ is a continuous, odd mapping and $0 \notin \psi(\partial D)$. Furthermore, for $x \in \partial D$ we have

$$\frac{\psi(x)}{|\psi(x)|} = \frac{\phi(x) - \phi(-x)}{|\phi(x) - \phi(-x)|} \neq \frac{\phi(-x) - \phi(x)}{|\phi(x) - \phi(-x)|} = \frac{\psi(-x)}{\psi(-x)}.$$

Set
$$H(x, t) := \phi(x) - t\phi(-x)$$

for every $t \in [0, 1]$ and every $x \in \bar{D}$. Clearly, $H$ is a $C^0$ homotopy between $\phi$ and $\psi$. If $H(x, t) = 0$ for some $x \in \partial D$ and some $t \in [0, 1]$, we have $\phi(x) = t\phi(-x)$ and

$t > 0$. This implies that $\frac{\phi(x)}{|\phi(x)|} = \frac{\phi(-x)}{|\phi(-x)|}$, which yields a contradiction. Therefore, $H(x,t) \neq 0$ for every $x \in \partial D$ and for every $t \in [0,1]$, and we conclude that

$$d(\phi, D, 0) = d(\psi, D, 0).$$

Due to the first step, in the sequel we assume, in addition, that $\phi$ is an odd mapping.

*Second step.* We prove that we can assume, without loss of generality, that $\phi(x) \equiv x$ in a neighbourhood of 0. Let $\epsilon > 0$ be such that $U := \bar{B}(0, \epsilon) \subset D$ and set $D_1 := D \setminus U$. We define $\phi_1 : U \cup \partial D \to \mathbb{R}^N$ by

$$\phi_1(x) := \begin{cases} x & x \in U \\ \phi(x) & x \in \partial D. \end{cases}$$

Then $\phi_1$ is a continuous function,

$$\partial D_1 = \partial D \cup \partial U = \partial D \cup \partial B(0, \epsilon)$$

and $\phi_1$ is an odd mapping. Let

$$\phi_2 := \phi_1|_{\partial D_1}.$$

As $\phi_2$ is nowhere zero, by Lemma 3.21 there exists a continuous extension of $\phi_2$, $\phi_3 : \bar{D}_1 \to \mathbb{R}^N$, such that $\phi_3$ is odd and nowhere zero on $\bar{D}_1 \cap \mathbb{R}^{N-1}$. Define

$$\phi_4(x) := \begin{cases} \phi_3(x) & x \in \bar{D}_1 \\ x & x \in U. \end{cases}$$

We show that $\phi_4$ is well defined. Indeed, if $x \in U \cap \bar{D}_1 = U \cap \bar{D}$, then $x \in \partial U$ and

$$\phi_3(x) = \phi_2(x) = x.$$

It is also easy to show that $\phi_4$ is an odd, continuous mapping. For every $x \in \partial D$, we have $x \in \partial D_1$ and

$$\phi_4(x) = \phi_3(x) = \phi_2(x) = \phi_1(x) = \phi(x).$$

Therefore, $d(\phi_4, D, 0) = d(\phi, D, 0)$, $0 \notin \phi_4(\partial D)$ and $\frac{\phi_4(x)}{|\phi_4(x)|} \neq \frac{\phi_4(x)}{|\phi_4(-x)|}$ for every $x \in \partial D$.

*Third step.* We show that $d(\phi_3, D_1, 0)$ is an even number. Let $K := \bar{D}_1 \cap \mathbb{R}^{N-1}$. Then $0 \notin \phi_3(K)$ and it is easy to see that $K$ is a compact set included in $\bar{D}_1$. By the excision property of the degree (see Theorem 2.7 (2)) we have

$$d(\phi_3, D_1, 0) = d(\phi_3, D_1 \setminus K, 0) = d(\phi_3, D_1^+, \cup D_1^-, 0),$$

where $D_1^+ := D_1 \cap \mathbb{R}_+^N$ and $D_1^- := D_1 \cap \mathbb{R}_-^N$. By the decomposition property of the degree (see Theorem 2.7 (1)) we deduce that

$$d(\phi_3, D_1, 0) = d(\phi_3, D_1^+, 0) + d(\phi_3, D_1^-, 0)$$

and, using the invariance of the degree under a $C^1$ change of variables (see Theorem 1.20), we have that

$$d(\phi_3, D_1^-, 0) = d((-I) \circ \phi_3 \circ (-I), -I(D_1^+), -I(0)) = d(\phi_3, D_1^+, 0).$$

Therefore, $d(\phi_3, D_1, 0) = 2d(\phi_3, D_1^+, 0)$.

*Fourth step.* We conclude that $d(\phi_4, D, 0)$ is an odd number. Since $\phi|_{\partial D} = \phi_4|_{\partial D}$, $\phi_4|_{\partial \text{int} U} = I$, $\phi_4|_{D_1} = \phi_3|_{D_1}$, and $D \setminus U = D_1$, we have by the excision and decomposition properties of the degree (see Theorem 2.7),

$$
\begin{aligned}
d(\phi, D, 0) &= d(\phi_4, D, 0) \\
&= d(\phi_4, D \setminus \partial B(0, \epsilon), 0) \\
&= d(\phi_4, \text{int}\, U \cup (D \setminus U), 0) \\
&= d(\phi_4, \text{int}\, U, 0) + d(\phi_4, D \setminus U, 0) \\
&= d(\phi_4, \text{int}\, U \cup (D \setminus U), 0) \\
&= 1 + d(\phi_4, D \setminus U, 0) \\
&= 1 + d(\phi_3, D_1, 0) \\
&= 1 + 2d(\phi_3, D_1^+, 0).
\end{aligned}
$$

$\square$

**Theorem 3.25 [The Borsuk Fixed Point Theorem]** *Let $D \subset \mathbb{R}^N$ be an open, bounded, symmetric set containing $0$ and let $\phi : \partial D \to \mathbb{R}^M$ be a continuous function, $M < N$. Then there exists $x \in \partial D$ such that $\phi(x) = \phi(-x)$.*

**Proof** We identify $\mathbb{R}^M$ with $\{x = (x_1, \ldots, x_N) \in \mathbb{R}^N : x_{N+1} = \ldots x_N = 0\}$. Define $\psi(x) := \phi(x) - \phi(-x)$ for $x \in \partial D$. We must show that there exists $x \in \partial D$ such that $\psi(x) = 0$.

Assume, on the contrary, that $\psi(x) \neq 0$ for every $x \in \partial D$. Since $\psi : \partial D \to \mathbb{R}^M$ is a continuous, odd, nowhere zero mapping and

$$\frac{\psi(x)}{|\psi(x)|} \neq \frac{\psi(-x)}{|\psi(-x)|} \quad \text{for every } x \in \partial D,$$

we obtain by Theorem 3.23 that $d(\psi, D, 0)$ is an odd number. By Theorem 2.3 there exists $\epsilon_0 > 0$ such that, for every $0 < \epsilon < \epsilon_0$,

$$d(\psi, D, p_\epsilon) = d(\psi, D, 0),$$

where

$$p_\epsilon = (0, \ldots, 0, \epsilon).$$

Therefore, $d(\psi, D, p_\epsilon) \neq 0$ and, by Theorem 2.1, $p_\epsilon \in \psi(D) \subset \mathbb{R}^M$ for every $0 < \epsilon < \epsilon_0$, which contradicts $p_\epsilon \notin \mathbb{R}^M$. $\square$

The following corollary is also known as the Borsuk–Ulam Theorem.

**Corollary 3.26 [Borsuk–Ulam Theorem]** *Let $S^N \subset \mathbb{R}^{N+1}$ be the N-sphere and let $\phi : S^N \to \mathbb{R}^N$ be a continuous mapping. Then there exists $x \in S^N$ such that $\phi(-x) = \phi(x)$.*

**Proof** The result follows immediately from Theorem 3.25, where we set $D := B(0,1) \subset \mathbb{R}^{N+1}$. $\qquad\square$

**Corollary 3.27** *Let $S^N$ be covered by N closed subsets $A_1, \ldots, A_N$. Then some $A_i$ must contain a pair of antipodal points, i.e. there exists a set $A_i$ and there exists $x \in A_i$ such that $-x \in A_i$.*

**Proof** For $x \in S^N$ we define

$$f_i(x) := \rho(x, A_i)$$

and

$$f(x) := (f_1(x), \ldots, f_N(x)),$$

and we recall that $\rho(x, y) = \max\{|x_j - y_j| : j = 1, \ldots, N\}$. Clearly, the function $f : S^N \to \mathbb{R}^N$ is continuous and by the Borsuk–Ulam Theorem (Corollary 3.26) there exists $\xi \in S^N$ such that $f(\xi) = f(-\xi)$. Since $\{A_i\}_i$ cover $S^N$, there exists some $i \in \{1, \ldots, N\}$ such that $\xi \in A_i$ and so $f_i(\xi) = 0 = f_i(-\xi)$. Since $A_i$ is closed, we conclude that $\xi, -\xi \in A_i$. $\qquad\square$

**Theorem 3.28 [The Ham-Sandwich Theorem]** *Let $X_1, \ldots, X_N \subset \mathbb{R}^N$ be N bounded, measurable sets. Then there exists an (N-1)-hyperplane $Y \subset \mathbb{R}^N$ that bisects each of the sets $X_1, \ldots, X_N$.*

**Proof** Without loss of generality, we assume that

$$\mathcal{L}^N(X_i) > 0, \quad i = 1, \ldots, N. \tag{3.1}$$

Recall that an (N-1)-hyperplane $\Pi_{b,\alpha} \subset \mathbb{R}^N$ has the form

$$\Pi_{b,\alpha} = \left\{(x_1, \ldots, x_N) \in \mathbb{R}^N : x_1 b_1 + \ldots + x_N b_N = \alpha\right\}$$

where $b \in \mathbb{R}^N$, $b \neq 0$, and $\alpha \in \mathbb{R}$. We identify $\mathbb{R}^N$ with $\{(y_1, \ldots, y_{N+1}) \in \mathbb{R}^{N+1} : y_{N+1} = 0\}$. Let $c = (0, \ldots, 0, 1) \in \mathbb{R}^{N+1}$ and define $f : S^N \to \mathbb{R}^N$ by

$$f := (f_1, \ldots, f_N), \quad f_i(x) := \mathcal{L}^N\left(\{y \in X_i : (y - c) \cdot x \geq 0\}\right).$$

We show that $f_i$ is continuous. Fix $x \in S^N$ and let $\{x_r\} \subset \mathbb{R}^N$ be a sequence such that $x_r \to x$ when $n$ tends to infinity. By Lebesgue's Dominated Convergence Theorem and since $X_i$ is a bounded set, we have

$$\chi_{\{y \in X_i \,:\, (y-c) \cdot x_r \geq 0\}} \to \chi_{\{y \in X_i \,:\, (y-c) \cdot x \geq 0\}}$$

almost everywhere and so $f_i(x_r) \to f_i(x)$. By the Borsuk–Ulam Theorem there exists $a \in S^N$ such that

$$f(a) = f(-a),$$

i.e. for every $i = 1, \dots, N$,

$$\mathcal{L}^N \left( \{ y \in X_i \ : \ (y - c) \cdot a \geq 0 \} \right) = \mathcal{L}^N \left( \{ y \in X_i \ : \ (y - c) \cdot a \leq 0 \} \right). \quad (3.2)$$

We claim that there exists $j \in \{1, \dots, N\}$ such that $a_j \neq 0$. Assume, on the contrary, that $a_1 = \dots = a_N = 0$. Since $a \in S^N$ we have $a_{N+1} = \pm 1$ and (3.2) implies that

$$\mathcal{L}^N \left( \{ y \in X_i \ : \ y_{N+1} - 1 \geq 0 \} \right) = \mathcal{L}^N \left( \{ y \in X_i \ : \ y_{N+1} - 1 \leq 0) \} \right).$$

As $X_i \subset \{ y \in \mathbb{R}^{N+1} \ : \ y_{N+1} = 0 \}$, we obtain $\mathcal{L}^N(X_i) = 0$, contradicting (3.1). Now (3.2) is equivalent to

$$\mathcal{L}^N \left( \{ (y_1, \dots, y_N) \in X_i \ : \ a \cdot y \geq a_{N+1} \} \right)$$
$$= \mathcal{L}^N \left( \{ (y_1, \dots, y_N) \in X_i \ : \ a \cdot y \leq a_{N+1} \} \right)$$

where $a \cdot y = a_1 y_1 + \dots + a_N y_N$. Setting

$$Y := \left\{ (y_1, \dots, y_N) \in \mathbb{R}^N : a \cdot y = a_{N+1} \right\},$$

we conclude that $Y$ bisects each of $X_1, \dots, X_N$. $\qquad \square$

## 3.3   The Jordan Separation Theorem

In $\mathbb{R}^2$ the Jordan Separation Theorem asserts that if $C$ is a Jordan curve, then $\mathbb{R}^2 \setminus C$ is a union of two connected, disjoint, open sets $D_1$ and $D_2$ such that $D_1$ is bounded and $D_2$ is unbounded. In this section we present a generalization of this theorem in $\mathbb{R}^N$. We recall that $K, L \subset \mathbb{R}^N$ are said to be *homeomorphic* if there exists a bijection $h : K \to L$ such that $h$ and $h^{-1}$ are both continuous.

**Theorem 3.29 [Jordan Separation Theorem]** *Let $K, L \subset \mathbb{R}^N$ be two compact sets such that $K$ and $L$ are homeomorphic. Then either $K^c$ and $L^c$ have the same finite number of connected components or both have countably infinitely many connected components.*

**Proof** Let $\{ \Delta_i : i \in I \}$ be the family of connected components of $K^c$. The sets $\Delta_i$ are mutually disjoint and, since $K$ is closed, each $\Delta_i$ is open. Therefore, there are at most countably infinitely many $\Delta_i$, since each of them contains a point with rational co-ordinates. We may write $I \subset \mathbb{N}$ and assume without loss of generality that if $i \in I$ and $i \geq 2$, then $i - 1 \in I$. As $K$ is a bounded set, exactly one of the connected components is unbounded, say $\Delta_0$. By the same argument, let $\{ D_r : r \in R \}$ be the set of connected components of $L^c$ such that $R \subset \mathbb{N}$, and $R$ has the property that $r \in R$ and $r \geq 2$ implies that $r - 1 \in R$.

Let $D_0$ be the unique, unbounded, connected component. Let $h : K \to L$ be a homeomorphism. By the Tietze Extension Theorem (Theorem 1.15) there exist a continuous extension of $h$, $\phi : \mathbb{R}^N \to \mathbb{R}^N$, and a continuous extension of $h^{-1}, \psi : \mathbb{R}^N \to \mathbb{R}^N$. We show that, for every $i, j \geq 1$,

$$\delta_{ij} = \sum_{k=1}^{|R|} d(\phi, \Delta_i, D_k) d(\psi, D_k, \Delta_j),$$

$$\delta_{ij} = \sum_{k=1}^{|I|} d(\phi, \Delta_k, D_i) d(\psi, D_j, \Delta_k)$$

and conclude the theorem by means of an elementary argument of linear algebra. Here $\delta_{ij}$ denotes the Kronecker symbol, equal to zero if $i \neq j$ and one if $i \neq j$. Fix $j \in \mathbb{N}$ and let $\{G_l^j : l \in \Lambda\}$ be the set of connected components of $(\phi(\partial\Delta_j))^c$, where $\Lambda \subset \mathbb{N}$ and $\Lambda$ has the property that $l \in \Lambda$ and $l \geq 2$ implies $l - 1 \in \Lambda$. Since $\partial\Delta_j \subset K$ and is bounded, then $\phi(\partial\Delta_j)$ is a compact set and so exactly one of the $G_l^j$ is unbounded, say $G_0^j$. We have

$$K^c = \cup_{j \in I} \Delta_j, \ L^c = \cup_{r \in R} D_r, \ (\phi(\partial\Delta_j))^c = \cup_{l \in L} G_l^j.$$

*Claim 1.* For every $r \in \mathbb{N}$, there exists $l \in \mathbb{N}$ such that $D_r \subset G_l^j$.

Indeed, fixing $r \in \mathbb{N}$, we observe that $\partial\Delta_j \subset K$ implies that $\phi(\partial\Delta_j) \subset \phi(K) = L$ and so $L^c \subset \phi(\partial\Delta_j)^c$. Therefore, $D_r \subset L^c \subset \phi(\partial\Delta_j)^c$ and, since $D_r$ is connected, there exists $l \in \mathbb{N}$ such that $D_r \subset G_l^J$. The collection $\{D_r\}$ may be relabelled $\{D_{l,k}^j\}$ in such a way that

$$U_0^j = D_0 \cup D_{0,1}^j \cup D_{0,2}^j \cup \ldots \subset G_0^j,$$

$$U_l^j = D_{l,1}^j \cup D_{l,2}^j \cup \ldots \subset G_l^j.$$

*Claim 2.* $d(\psi, G_l^j, p) = \sum_{k=1}^{+\infty} d(\psi, D_{l,k}^j, p)$ for $p \in \Delta_i$ and $l \geq 1$.

First we prove that $d(\psi, G_l^j, p)$ is well defined. It suffices to prove that $\partial G_l^j \subset L$. Let $x \in \partial G_l^j$ and assume that $x \notin L$. We obtain that $x \in D_r$ for a suitable $r$ and so, since $D_r \subset G_{l(r)}^j$, we conclude that $x \in G_{l(r)}^j$. Thus $G_l^j \cap G_{l(r)}^j \neq \emptyset$ and so $l = l(r)$. Therefore, we have $x \in \partial G_l^j \cap G_l^j$, which yields a contradiction. Hence $\partial G_l^j \subset L$ and

$$\psi(\partial G_l^j) \subset \psi(L) = K.$$

Finally, $p \in \Delta_i$ implies that $p \notin K$, which, in turn, implies that $p \notin \psi(\partial G_l^j)$ and so $d(\psi, G_l^j, p)$ is well defined.

Since $\partial D_{l,k}^j = \partial D_r$ for a suitable $r$ and $L^c = \cup_{r \in R} D_r$, we deduce that $\partial D_{l,k}^j \subset L$. Hence $d(\psi, D_{l,k}^j, p)$ is well defined.

Consider the compact set $M := \bar{G}_l^j \setminus U_l^j$ and fix $x \in M$. Then either $x \in \partial G_l^j$ or $x \in G_l^j \setminus U_l^j$; therefore $x \notin D_r$ for every $r \in \mathbb{N}$. Thus $x \in L$ and $\psi(x) \in \psi(L) = K$. Moreover, if $p \in \Delta_i$, then $p \notin \psi(M)$ and by the excision property and the decomposition property of the domain (see Theorem 2.7) we obtain that

$$d(\psi, G_l^j, p) = d(\psi, G_l^j \setminus M, p) = d(\psi, U_l^j, p) = \sum_{k=1}^{+\infty} d(\psi, D_{l,k}^j, p).$$

*Claim 3.* We claim that $\delta_{ij} = \sum_{k=1}^{+\infty} d(\psi, D_k, \Delta_i) d(\phi, \Delta_j, D_k)$ for $i, j \geq 1$.

Let us recall that $d(\psi, D_k, \Delta_i) = d(\phi, \Delta_k, p)$ for all $p \in \Delta_i$ and these are well defined since $\Delta_i \subset K^c = (\phi(L))^c$. Fixing $p \in \Delta_i$ and using the multiplication theorem, we have

$$d(\psi \circ \phi, \Delta_j, p) = \sum_{l=1}^{+\infty} d(\psi, G_l^j, p) d(\psi, \Delta_j, G_l^j)$$

and the summation is finite. Since $D_{l,k}^j \subset G_l^j$, we deduce that $d(\phi, \Delta_j, G_l^j) = d(\phi, \Delta_j, D_{l,k}^j)$ for every $k$, which, together with claim 2, yields

$$
\begin{aligned}
d(\psi \circ \phi, \Delta_j, p) &= \sum_{l=1}^{+\infty} d(\psi, G_l^j, p) d(\psi, \Delta_j, G_l^j) \\
&= \sum_{l=1}^{+\infty} \sum_{k=1}^{+\infty} d(\psi, D_{l,k}^j, p) d(\psi, \Delta_j, D_{l,k}^j).
\end{aligned}
$$

Recall that $d(\psi, \Delta_j, D_0^j k) = d(\phi, \Delta_j, G_0^j) = 0$ because $G_0^j$ is unbounded and so

$$
\begin{aligned}
d(\psi \circ \phi, \Delta_j, p) &= \sum_{l=0}^{+\infty} \sum_{k=1}^{+\infty} d(\psi, D_{l,k}^j, p) d(\phi, \Delta_j, D_{l,k}^j) \\
&= \sum_{r=1}^{+\infty} d(\psi, D_r, p) d(\phi, \Delta_j, D_r),
\end{aligned}
$$

since $\{D_r : r \geq 1\} = \{D_{l,k}^j : l \geq 0, k \geq 1\}$. The last equality can be written as

$$d(\psi \circ \phi, \Delta_j, \Delta_i) = \sum_{r=1}^{+\infty} d(\psi, D_r, \Delta_i) d(\phi, \Delta_j, D_r).$$

Using the fact that $\partial \Delta_j \subset K$ and $\psi \circ \phi(x) = x$ for every $x \in K$, we have

$$d(\psi \circ \phi, \Delta_j, \Delta_i) = d(I, \Delta_j, \Delta_i) = \delta_{ij};$$

thus

$$\delta_{ij} = \sum_{r=1}^{+\infty} d(\psi, D_r, \Delta_i) d(\phi, \Delta_j, D_r). \tag{3.3}$$

Using a similar argument we have that

$$\delta_{ij} = \sum_{r=1}^{+\infty} d(\psi, D_i, \Delta_r) d(\phi, \Delta_r, D_j). \tag{3.4}$$

*Claim 4.* $\{D_r : r \geq 1, r \in R\}$ and $\{\Delta_j : j \geq 1, j \in I\}$ are in bijection.

We divide the proof of claim 4 in three cases.

*First case.* $R < +\infty$ and $I < +\infty$.

Let $A \in \mathbb{R}^{R \times I}$ be defined by

$$a_{rs} := d(\phi, \Delta_r, D_s)$$

and let $B \in \mathbb{R}^{I \times R}$ be defined by

$$b_{rs} := d(\psi, D_r, \Delta_s).$$

We have $AB \in \mathbb{R}^{R \times R}$ and

$$(AB)_{ij} = \sum_{k=1}^{I} a_{ik} b_{kj} = \sum_{k=1}^{I} d(\phi, \Delta_i, D_k) d(\psi, D_k, \Delta_j) = \delta_{ij};$$

hence $AB$ is the identity matrix of $\mathbb{R}^{R \times R}$ and so it has rank $R$. We observe that

$$R = \text{rank } (AB) \leq \text{rank } A \leq \min\{R, I\} \leq I.$$

Similarly, $BA \in \mathbb{R}^{I \times I}$ is the identity matrix of $\mathbb{R}^{I \times I}$,

$$I = \text{rank } (BA) \leq \text{rank } B \leq \min\{R, I\} \leq R.$$

and we conclude that $R = I$.

*Second case.* $R = +\infty$ and $I = +\infty$.

In this case, it is obvious that $\{D_r : r \in R\}$ and $\{\Delta_i : i \in I\}$ are in bijection.

*Third case.* $R = +\infty$ and $I < +\infty$.

Let

$$X = \mathbb{R}[x]$$

be the set of real polynomials with the basis $1, x, x^2, x^3, \ldots$. Set

$$Y := \text{span } \{1, x, x^2, \ldots, x^I\}$$

and define two linear applications $f : Y \rightarrow X$ and $g : X \rightarrow Y$ by

$$f(x^j) := \sum_{r=1}^{\infty} a_{rj} x^r, j = 1, \ldots, I$$

and

$$g(x^j) := \sum_{r=1}^{\infty} a_{rj} x^r, j = 1, \ldots, \infty.$$

Since $a_{rj} = 0$, except for finitely many $r \in \mathbb{N}$, $j \in \{1, \ldots, I\}$, $f$ and $g$ are well defined. By (3.3) and (3.4), we obtain that

$$f \circ g = I_X$$

and

$$g \circ f = I_Y,$$

where $I_X$ stands for the identity matrix of $X$ and $I_Y$ stands for the identity matrix of $X$. Hence

$$+\infty = \text{rank } (f \circ g) \leq \text{rank } f \leq I,$$

which yields a contradiction.

*Fourth case.* $R < +\infty$ and $I = +\infty$. The proof of this case is identical to the proof of the third case. $\qquad \square$

We give an application of the Jordan Separation Theorem.

**Theorem 3.30 [Invariance of Domain]** *Let $D \subset \mathbb{R}^N$ be an open set and let $\phi : D \to \mathbb{R}^N$ be a continuous, injective function. Then $\phi(D)$ is an open set.*

**Proof** We observe that for every $p \in D$ there exists $r(p) > 0$ such that $B(p, r(p)) \subset\subset D$. Thus

$$\phi(D) = \bigcup_{p \in D} \phi(B(p, r(p))).$$

To show that $\phi(D)$ is an open set, it suffices to show that $\phi(B(p, r(p)))$ is an open set for every $p \in D$. Fix $p \in D$, and set $B := B(p, r(p))$. Since $\partial B$ is a compact set and $\phi$ is injective, we obtain that $\phi : \partial B \to \phi(\partial B)$ is a homeomorphism and, by the Jordan Separation Theorem, $(\phi(\partial B))^c$ has two connected components, $\Delta_1$ open and bounded and $\Delta_2$ open and unbounded.

We claim that $\phi(B) = \Delta_1$. Indeed, as $\bar{B}$ is a compact set and $\phi$ is injective, we have that $\phi|_{\bar{B}} : \bar{B} \to \phi(\bar{B})$ is a homeomorphism and so, by the Jordan Separation Theorem, we conclude that $(\phi(\bar{B}))^c$ is an unbounded, connected set. Considering, in addition, the fact that $(\phi(\bar{B}))^c \subset (\phi(\partial B))^c$, we deduce that either $(\phi(\bar{B}))^c \subset \Delta_1$ or $(\phi(\bar{B}))^c \subset \Delta_2$. Since $\Delta_1$ is bounded and $(\phi(\bar{B}))^c$ is not, we must have

$$(\phi(\bar{B}))^c \subset \Delta_2.$$

This implies that
$$\Delta_1 \cup \phi(\partial B) = \Delta_2^c \subset \phi(B) \cup \phi(\partial B)$$
and so, as $\phi$ is an injection, we obtain that

$$\Delta_1 \subset \phi(B). \tag{3.5}$$

Similarly, $\phi(B)$ is a connected, bounded set, $\phi(B) \subset (\phi(\partial B))^c$, and we conclude that
$$\phi(B) \subset \Delta_1 \text{ or } \phi(B) \subset \Delta_2,$$
which, together with (3.5), yields

$$\phi(B) = \Delta_1.$$

We conclude that $\phi(B)$ is an open set.                    □

**Corollary 3.31** *Let $M < N$ be two integer numbers. Then there is no injective mapping $\phi : \mathbb{R}^N \to \mathbb{R}^M$.*

**Proof**  Assume that there exists an injective mapping $\phi : \mathbb{R}^N \to \mathbb{R}^M$ and define $\psi : \mathbb{R}^N \to \mathbb{R}^N$ by
$$\psi(x) := (\phi(x), a) \quad x \in \mathbb{R}^N,$$
where $a = (0, \ldots, 0) \in \mathbb{R}^{N-M}$. Then $\psi$ is injective and, by Theorem 3.30, $\psi(\mathbb{R}^N) = \mathbb{R}^M \times \{a\}$ is an open set of $\mathbb{R}^N$. This yields a contradiction.                    □

**Corollary 3.32** *Let $N < M$ be two integer numbers and let $\phi : \mathbb{R}^N \to \mathbb{R}^M$ be an injective mapping. Then $(\phi(\mathbb{R}^N))^c$ is dense in $\mathbb{R}^M$.*

The proof of this result uses Baire's Theorem. We recall its statement and refer the reader to Eisenberg (1974) for a proof.

**Theorem 3.33 [Baire's Theorem]** *Let $X$ be a complete, metric space and let $\{U_i, i \in \mathbb{N}\}$ be a collection of open, dense subsets of $X$. Then $\cap_{i \in \mathbb{N}} U_i$ is dense in $X$.*

**Proof of Corollary 3.32.**  We first prove that $(\phi(I_k^N))^c$ is dense in $\mathbb{R}^M$, where $I_k := [-k, k]$.

Assume, on the contrary, that there exist $a \in \mathbb{R}^M$ and $r > 0$ such that $\bar{B}(a, r) \subset \phi(I_k^N)$. Setting
$$K := \phi^{-1}(\bar{B}(a, r)),$$
we observe that $K$ is a compact set and so, since $\phi$ is injective,

$$g : K \to \bar{B}(a, r)$$

is a homeomorphism, where
$$g := \phi|_K.$$

Then

$$g^{-1} : \bar{B}(a,r) \subset \mathbb{R}^M \to K \subset \mathbb{R}^N$$

is a homeomorphism, contradicting the Borsuk–Ulam Theorem (see Corollary 3.26).

Next we show that $(\phi(\mathbb{R}^N))^c$ is dense in $\mathbb{R}^M$.

Since $\mathbb{R}^N = \cup_{k \in \mathbb{N}} I_k^N$, we obtain

$$(\phi(\mathbb{R}^N))^c = \underset{k \in \mathbb{N}}{\cap} (\phi(I_k^N))^c.$$

We observe that $I_k^N$ are compact sets, $\phi$ is continuous, and so $\phi(I_k^N)$ are compact sets. Hence $(\phi(I_k^N))^c$ are open sets, each $(\phi(I_k^N))^c$ is dense in $\mathbb{R}^M$, and thus by Baire's Theorem $\cap_{k \in \mathbb{N}}(\phi(I_k^N))^c$ is dense in $\mathbb{R}^M$. $\qquad\square$

**Remark 3.34** Peano's Theorem states that there exists a continuous surjection $f : [0,1] \to [0,1] \times [0,1]$. We refer the reader to Eisenberg (1974, pp.367–370) or Kuratowsky (1966, pp. 150–151). Actually, it is possible to prove that, for every $N, M \geq 1$, there exists a continuous surjection $\phi : \mathbb{R}^N \to \mathbb{R}^M$ (see Exercise 3.3).

**Theorem 3.35** Let $D \subset \mathbb{R}^N$ be an open, bounded set and let $\phi \in C(\bar{D})^N$ be an injective mapping. Then for every $p \in \phi(D)$ we have

$$d(\phi, D, p) = \pm 1.$$

**Proof** By Theorem 3.30, $\phi(D)$ is an open set and $\phi^{-1} : \phi(D) \to D$ is a homeomorphism. Let $p \in \phi(D)$. Then there exists $r > 0$ such that $B := B(p,r) \subset\subset \phi(D)$ and we set

$$\Delta := D \setminus \phi^{-1}(\partial B).$$

We want to apply the Multiplication Theorem to $\phi \circ \phi^{-1} : \bar{B} \to \phi(D)$ and $\phi^{-1} : \bar{B} \to D$. Let $\{\Delta_j : j \in J\}$ be the countable family of the connected components of $\Delta$. We have

$$p \notin \phi \circ \phi^{-1}(\partial B) = \partial B$$

and also,

$$p \notin \phi(\partial D).$$

Therefore, by the Multiplication Theorem (Theorem 2.10),

$$1 = d(I, B, p) = d(\phi \circ \phi^{-1}, B, p) = \sum_{i=1}^{+\infty} d(\phi, \Delta_i, p) d(\phi^{-1}, B, \Delta_i). \qquad (3.6)$$

Also, by the Jordan Separation Theorem, $\phi^{-1}(\partial B)^c$ has two open connected components $D_1$ and $D_2$ such that $D_1$ is bounded and $D_2$ is unbounded.

We claim that there exists $i_0 \in J$ such that $D_1 = \Delta_{i_0}$. Indeed, $\Delta = D \setminus \phi^{-1}(\partial B) \subset \phi^{-1}(\partial B)^c$ and, by an argument similar to that of the proof of Theorem 3.30, we have that

$$D_1 = \phi^{-1}(B) \subset D \setminus \phi^{-1}(\partial B) = \Delta = \cup_{j \in J} \Delta_j.$$

Therefore, $D_1 \subset \Delta_{i_0}$ for some $i_0 \in J$ and as

$$\Delta_{i_0} \subset \Delta \subset \phi^{-1}(\partial B)^c \subset D_1 \cup D_2,$$

we have $\Delta_{i_0} \subset D_1$ and we conclude that

$$\Delta_{i_0} = D_1.$$

Hence $\Delta_i \subset D_2$ for every $i \neq i_0$ and so

$$d(\phi^{-1}, B, \Delta_i) = d(\phi^{-1}, B, D_2) = 0 \text{ for every } i \neq i_0,$$

which, together with (3.6), yields

$$1 = d(\phi, \Delta_{i_0}, p) d(\phi^{-1}, B, \Delta_{i_0})$$

and so

$$d(\phi, \Delta_{i_0}, p) = \pm 1. \tag{3.7}$$

Now it suffices to show that $d(\phi, \Delta_{i_0}, p) = d(\phi, D, p)$. Let

$$K := \bar{D} \setminus \Delta_{i_0}.$$

Since $d(\phi, \Delta_{i_0}, p) \neq 0$, by Theorem 2.1 we deduce that $p \in \phi(\Delta_{i_0})$ and, as $\phi$ is injective, we deduce that

$$p \notin \phi(\bar{D} \setminus \Delta_{i_0}).$$

Using the excision property of the degree (see Theorem 2.7), we have

$$d(\phi, \Delta_{i_0}, p) = d(\phi, D \setminus K, p) = d(\phi, D, p),$$

which, together with (3.7), yields

$$d(\phi, D, p) = \pm 1.$$

□

## 3.4 Exercises

**Exercise 3.1 [Perron–Frobenius Theorem]** Let $A = (a_{ij})$ be an $N \times N$ matrix such that $a_{ij} \geq 0$ for all $i, j$. Prove that there exist $\lambda \geq 0$, $x \neq 0$ such that $x_i \geq 0$ for every $i$ and $Ax = \lambda x$.

*Solution 3.1.* Let $D := \{x \in \mathbb{R}^N : x_i \geq 0, i = 1, \ldots, N, \ \sum_{i=1}^{N} x_i = 1\}$. If $Ax = 0$ for some $x \in D$, then it suffices to set $\lambda = 0$. Assume that $Ax \neq 0$ for every $x \in D$. Then $\sum_{i=1}^{N}(Ax)_i \geq \alpha$ for every $x \in D$ and for some $\alpha > 0$, and the function $f : D \to \mathbb{R}^N$ defined by

$$f(x) := \frac{Ax}{\sum_{i=1}^{N}(Ax)_i}$$

is continuous in $D$. It is clear that

$$\sum_{i=1}^{N} f_i(x) = 1$$

and

$$f_i(x) \geq 0,$$

for every $x \in D$. Hence,

$$f(D) \subset D.$$

By the Brouwer Fixed Point Theorem (see Corollary 3.8), there exists $x_0 \in D$ such that $f(x_0) = x_0$. Setting $\lambda := \sum_{i=1}^{N}(Ax_0)_i$ we obtain that

$$Ax_0 = \lambda x_0.$$

A generalization of this theorem can be found in Varga (1962).

**Exercise 3.2** Let $K, L \subset \mathbb{R}^N$ be two compact sets such that $K \subset L$. Assume that $f : K \to \mathbb{R}^M$ is a continuous function. Prove that there exists a continuous function $g : L \to \mathbb{R}^M$ which coincides with $f$ on $K$ such that $|f|_K = |g|_L = \sup\{|g_i|_L : i = 1, \ldots, M\}$.

*Solution 3.2.* Since $K$ is a compact set and $f : K \to \mathbb{R}^M$ is continuous, we deduce that $f$ is bounded. Let $|f_i|_K = \sup\{|f_i(x)| : x \in K\}$. We recall that $|f|_K = \sup\{|f_i| : i = 1, \ldots, M\}$. By Tietze's Extension Theorem, for each $i = 1, \ldots, m$ we obtain the existence of a continuous function $g_i : \mathbb{R}^N \to \mathbb{R}$ such that

$$g_i|K = f_i, \ \sup_{x \in L} g_i = \sup_{x \in K} f_i, \ \text{and} \ \inf_{x \in L} g_i = \inf_{x \in K} f_i.$$

We have $|g|_M \geq |g|_K = |f|_K$. Conversely $|g|_L = \max\{|g_i|_L : i = 1, \ldots, M\} = |g_{i_0}|_L$ for some $i_0 \in \{1, \ldots, M\}$ and, as $L$ is a compact set, we obtain that $|g_{i_0}|_L = |g_{i_0}(\bar{x})|$ for some $\bar{x} \in L$. There exist $a, b \in K$ such that

$$f_{i_0}(a) = \min_{x \in K} f_{i_0}(x) \leq g_{i_0}(\bar{x}) \leq \max_{x \in K} f_{i_0}(x) = f_{i_0}(b)$$

and either $|g_{i_0}(\bar{x})| \leq |f_{i_0}(b)|$ or $|g_i(\bar{x})| \leq |f_{i_0}(a)|$. This yields $|g|_L \leq |f|_K$ and so

$$|g|_L = |f|_K.$$

**Exercise 3.3** Assume that there exists a continuous surjection $f : [0,1] \to [0,1] \times [0,1]$ ([**Peano's Theorem**]). Prove that, for every $N, M \geq 1$, there exists a continuous surjection $\phi : \mathbb{R}^N \to \mathbb{R}^M$.

*Solution 3.3.* By Peano's Theorem, for every $k \in \mathbb{N}, k > 0$, there exists a continuous surjection $f_k : [2k - 2, 2k - 1] \to [-k, k] \times [-k, k]$. Define $g_k : [2k - 1, 2k] \to \mathbb{R}^2$ by

$$g_k(2k - 1 + t) := (1 - t)f_k(2k - 1) + tf_{k+1}(2k), \ t \in [0, 1].$$

We observe that $g_k$ is continuous and defining $\phi : \mathbb{R} \to \mathbb{R}^2$ as

$$\phi(x) := \begin{cases} f_1(0) & x \leq 0 \\ f_k(x) & x \in [2k - 2, 2k - 1] \\ g_k(x) & x \in [2k - 1, 2k], \end{cases}$$

then $\phi : \mathbb{R} \to \mathbb{R}^2$ is a continuous surjection. We easily deduce that, for every $N, M \geq 1$, there exists a continuous surjection $\phi : \mathbb{R}^N \to \mathbb{R}^M$.

**Exercise 3.4** Let $f : S^N \to S^N$ be a continuous function such that $f(x) \neq -x$ for every $x \in S^N$. Show that, if $N$ is even, then $f$ has a fixed point on $S^N$.

*Solution 3.4.* By Tietze's Extension Theorem (Theorem 1.15) there exists a continuous function $F : \mathbb{R}^{N+1} \to \mathbb{R}^{N+1}$ such that

$$F(x) = f(x) \neq -x \text{ for every } x \in S^N.$$

By Proposition 3.9, and as $N + 1$ is odd and $F(S^N) \subset S^N$, we conclude that there exist $\lambda > 0$ and $x_0 \in S^N$ such that

$$F(x_0) = \lambda x_0$$

i.e.

$$f(x_0) = \lambda x_0.$$

Since $|x_0| = 1 = |f(x_0)|$ and $f(x_0) \neq -x_0$, we must have

$$f(x_0) = x_0.$$

# 4

## MEASURE THEORY AND SOBOLEV SPACES

In this chapter we recall some properties of Sobolev functions and measure theory. In order to measure the sets on which a Sobolev function is continuous, we introduce the notion of Hausdorff measures and $p$-capacities. For a detailed description of these topics, we refer the reader to Evans and Gariepy (1992) and Ziemer (1989). The Hausdorff measures were first introduced by Carathéodory (1914). He developed only the Hausdorff linear measure in $\mathbb{R}^N$ and indicated how the $k$-dimensional measure could be defined for $k \in \mathbb{N}$. Later, motivated by the study of the dimension of the Cantor ternary set, Hausdorff (1919) developed the theory of the $k$-dimensional measure.

### 4.1  Review of measure theory

We recall some definitions and well-known results used in measure theory, such as the Riesz Representation Theorem, the Radon–Nikodym Differentiation Theorem and Vitali's Covering Theorem. For more details and for the proofs of the results presented in this section, we refer the reader to Evans and Gariepy (1992).

**Definition 4.1**

(i) *We say that $I \subset \mathbb{R}^N$ is a closed $N$-interval if there exist $a_i < b_i, i, \ldots, N$, such that*

$$I = [a_1, b_1] \times \ldots \times [a_N, b_N].$$

*We set $m(I) := (b_1 - a_1) \ldots (b_N - a_N)$.*

(ii) *We define the* Lebesgue outer measure $\mathcal{L}^N(E)$ *of a set $E \subset \mathbb{R}^N$ by*

$$\mathcal{L}^N(E) := \inf \left\{ \sum_{k=1}^{\infty} m(I_k) : E \subset \bigcup_{k \in \mathbb{N}} I_k, I_k \text{ closed } N\text{-interval} \right\}.$$

**Definition 4.2** *Let $X \subset \mathbb{R}^N$ be a nonempty set, let $\mathcal{P}(X)$ be the collection of the subsets of $X$, and let $\mu : \mathcal{P}(X) \to [0, \infty]$.*

(i) *$\mu$ is said to be an* outer measure *if*

$$\mu(\emptyset) = 0, \quad \mu(\bigcup_{i \in \mathbb{N}} A_i) \leq \sum_{i=1}^{\infty} \mu(A_i)$$

*for every $\{A_i\} \subset \mathcal{P}(X)$.*

(ii) $A \in \mathcal{P}(\mathcal{X})$ *is* measurable with respect to the measure $\mu$ *if*

$$\mu(B) = \mu(A \cap B) + \mu(B \setminus A)$$

*for every* $B \in \mathcal{P}(\mathcal{X})$.

(iii) *The* restriction *of the measure* $\mu$ *to some* $E \in \mathcal{P}(\mathcal{X})$, $\mu \lfloor E$, *is defined by*

$$\mu \lfloor E(A) := \mu(A \cap E),$$

*for every* $A \subset E$.

**Remark 4.3** Assume that $\mu$ is an outer measure on $X$. Then

(i) if $A \subset B \subset X$, then $\mu(A) \leq \mu(B)$;

(ii) if $A \subset X$ and $\mu(A) = 0$, then $A$ is measurable;

(iii) $A \subset X$ is measurable if and only if $X \setminus A$ is measurable;

(iv) the Lebesgue outer measure $\mathcal{L}^N$ is an outer measure; in this case, a $\mu$ measurable set is called, simply, a measurable set.

**Definition 4.4** *Let* $X \subset \mathbb{R}^N$ *be nonempty and let* $\mathcal{A}$ *be a collection of subsets of* $X$. *We say that* $\mathcal{A}$ *is a* $\sigma$ algebra *if*

(i) $\emptyset \in \mathcal{A}$;

(ii) $A \in \mathcal{A}$ *implies that* $X \setminus A \in \mathcal{A}$;

(iii) $\{A_k, k \in \mathbb{N}\} \subset \mathcal{A}$ *implies that* $\cup_{k \in \mathbb{N}} A_k \in \mathcal{A}$.

**Remark 4.5** If $\mu$ is an outer measure on $X$, then the collection of all $\mu$ measurable subsets of $X$ forms a $\sigma$ algebra.

**Definition 4.6** *If* $X = \mathbb{R}^N$, *then the* Borel $\sigma$ algebra *is the smallest* $\sigma$ *algebra containing the open subsets of* $\mathbb{R}^N$.

**Definition 4.7** *Let* $\mu$ *be an outer measure on* $\mathbb{R}^N$.

(i) $\mu$ *is said to be* regular *on* $\mathbb{R}^N$ *if for every* $A \subset \mathbb{R}^N$ *there exists a* $\mu$ *measurable set* $B \subset \mathbb{R}^N$ *such that* $A \subset B$ *and* $\mu(A) = \mu(B)$.

(ii) $\mu$ *is said to be a* Borel measure *on* $\mathbb{R}^N$ *if every Borel set is* $\mu$ *measurable.*

(iii) $\mu$ *is said to be a* Borel regular measure *on* $\mathbb{R}^N$ *if* $\mu$ *is a Borel measure and for every* $A \subset \mathbb{R}^N$ *there exists a Borel set* $B \subset \mathbb{R}^N$ *such that* $A \subset B$ *and* $\mu(A) = \mu(B)$.

(iv) $\mu$ *is said to be a* Radon measure *on* $\mathbb{R}^N$ *if* $\mu$ *is a Borel regular measure and* $\mu(K) < \infty$ *for every compact set* $K \subset \mathbb{R}^N$.

**Lemma 4.8** *Let* $\mu$ *be a Borel regular measure on* $\mathbb{R}^N$ *and let* $A \subset \mathbb{R}^N$ *be a* $\mu$ *measurable set such that* $\mu(A) < \infty$. *Then* $\mu \lfloor A$ *is a Radon measure on* $\mathbb{R}^N$.

**Definition 4.9** *Let* $X \subset \mathbb{R}^N$ *be nonempty, let* $\mu$ *be an outer measure on* $X$, *and let* $\mathcal{A}$ *be the* $\sigma$ *algebra of all* $\mu$ *measurable subsets of* $X$. *The restriction of* $\mu$ *to* $\mathcal{A}$ *is called* a measure on $X$.

**Definition 4.10** *Let $X \subset \mathbb{R}^N$ be nonempty and let $\mu$ be a measure on $X$. $A \subset X$ is said to be $\sigma$ finite if there exists a countable family $\{A_k, k \in \mathbb{N}\}$ of $\mu$ measurable subsets of $X$ such that*

$$A = \bigcup_{k \in \mathbb{N}} A_k$$

*and $\mu(A_k) < \infty$ for every $k \in \mathbb{N}$.*

**Definition 4.11** *Let $\mu$ be a Radon measure on $\mathbb{R}^N$ and let $A \subset \mathbb{R}^N$. We say that a property holds $\mu$ almost everywhere on $A$ if the property holds for every $x \in B$, for some $B \subset A$ such that $\mu(A \setminus B) = 0$.*

Let $\mu$ and $\nu$ be two Radon measures on $\mathbb{R}^N$. For each $x \in \mathbb{R}^N$ define

$$\overline{D_\mu \nu}(x) := \begin{cases} \limsup\limits_{r \to 0+} \dfrac{\nu(B(x,r))}{\mu(B(x,r))} & \mu(B(x,r)) > 0 \text{ for all } r > 0 \\ +\infty & \mu(B(x,r)) = 0 \text{ for some } r > 0 \end{cases}$$

and

$$\underline{D_\mu \nu}(x) := \begin{cases} \liminf\limits_{r \to 0+} \dfrac{\nu(B(x,r))}{\mu(B(x,r))} & \mu(B(x,r)) > 0 \text{ for all } r > 0 \\ +\infty & \mu(B(x,r)) = 0 \text{ for some } r > 0. \end{cases}$$

**Definition 4.12** *Let $\mu$ and $\nu$ be two Radon measures on $\mathbb{R}^N$ and let $x \in \mathbb{R}^N$. If $\overline{D_\mu \nu}(x) = \underline{D_\mu \nu}(x) < +\infty$, then we say that $\nu$ is differentiable with respect to $\mu$ and we write*

$$D_\mu \nu(x) = \overline{D_\mu \nu}(x) = \underline{D_\mu \nu}(x).$$

$D_\mu \nu(x)$ *is called the* derivative *of $\nu$ with respect to $\mu$.*

**Definition 4.13** *Let $\mu$ be a measure on $\mathbb{R}^N$. A function $f : \mathbb{R}^N \to [-\infty + \infty]$ is said to be $\mu$-measurable if $f^{-1}([-\infty, \alpha))$ is $\mu$-measurable for every $\alpha \in \mathbb{R}$. When $\mu$ is the Lebesgue measure, a $\mu$-measurable function is said to be* measurable.

**Definition 4.14** *Let $\mu$ be a measure on $\mathbb{R}^N$, let $1 \leq p \leq +\infty$, and let $f : \mathbb{R}^N \to [-\infty + \infty]$ be a $\mu$ -measurable function. We say that $f \in L^p(\mathbb{R}^N, \mu)$ if*

$$\int_{\mathbb{R}^N} |f|^p \, d\mu < +\infty \quad \text{for } p < +\infty,$$

*and if $p = +\infty$ there exists $M \in \mathbb{R}$ such that*

$$|f(x)| \leq M \quad \text{for } \mu \text{ a.e. } x \in \mathbb{R}^N.$$

The following is also known as the Radon–Nikodym Theorem.

**Theorem 4.15** *Let $\mu$ and $\nu$ be two Radon measures on $\mathbb{R}^N$. Then for $\mu$ almost every $x \in \mathbb{R}^N$, we have that $D_\mu \nu(x)$ exists and $D_\mu \nu(x) < \infty$. Furthermore, $D_\mu \nu$ is $\mu$-measurable.*

**Definition 4.16** *Let $\mu$ and $\nu$ be two measures on $\mathbb{R}^N$.*

(i) *We say that $\nu$ is* absolutely continuous *with respect to $\mu$, and we write*

$$\nu \ll \mu ,$$

*if $\mu(A) = 0$ implies $\nu(A) = 0$ for every $A \subset \mathbb{R}^N$.*

(ii) *We say that $\nu$ and $\mu$ are* mutually singular, *and we write*

$$\nu \perp \mu ,$$

*if there exists a Borel set $B \subset \mathbb{R}^N$ such that*

$$\nu(B) = \mu(\mathbb{R}^N \setminus B).$$

**Theorem 4.17 [Differentiation Theorem for Radon measures]** *Let $\mu$ and $\nu$ be two Radon measures on $\mathbb{R}^N$. Then $\nu = \nu_a + \nu_s$, where*

$$\nu_a \ll \mu, \; \nu_s \perp \mu.$$

*Moreover,*

$$D_\mu \nu = D_\mu \nu_a, \; D_\mu \nu_s = 0 \; \mu \text{ a.e.}$$

*and*

$$\nu(A) = \int_A D_\mu \nu \, d\mu + \nu_s(A)$$

*for every Borel set $A \subset \mathbb{R}^N$.*

**Theorem 4.18 [Lebesgue–Besicovitch Differentiation Theorem]** *Let $\mu$ be a Radon measure on $\mathbb{R}^N$, $1 \leq p < \infty$, and let $f \in L^p_{loc}(\mathbb{R}^N, \mu)$. Then*

$$\lim_{\epsilon \to 0+} \frac{1}{\mu(B(x, \epsilon))} \int_{B(x, \epsilon)} |f(y) - f(x)|^p \, d\mu(y) = 0 \qquad (4.1)$$

*for $\mu$ a.e. $x \in \mathbb{R}^N$.*

**Definition 4.19** *A point $x$ for which (4.1) holds is called a $p$-Lebesgue point of $f$ with respect to $\mu$. We say that a 1-Lebesgue point of $f$ with respect to $\mathcal{L}^N$ is a Lebesgue point of $f$.*

**Theorem 4.20 [Riesz Representation Theorem]** *Let $L : C_c(\mathbb{R}^N)^M \to \mathbb{R}$ be a linear functional such that*

$$\sup\{L(f) : f \in C_c(\mathbb{R}^N)^M, |f|_2 \leq 1, \operatorname{spt}(f) \subset K\} < \infty$$

*for each compact set $K \subset \mathbb{R}^N$. Then there exists a Radon measure $\mu$ on $\mathbb{R}^N$ and a $\mu$ measurable function $\sigma : \mathbb{R}^N \to \mathbb{R}^M$ such that*

(i) *$|\sigma(x)|_2 = 1$ for $\mu$ a.e. $x \in \mathbb{R}^N$,*

(ii) *$L(f) = \int_{\mathbb{R}^N} f \cdot \sigma \, d\mu$, for all $f \in C_c(\mathbb{R}^N)^M$.*

## 4.2  Hausdorff measures

The Hausdorff measure $H^1(E)$ of a set $E \subset \mathbb{R}^2$ was introduced for the purpose of generalizing the notion of length in the case where $E$ is a smooth curve. As a first approximation, we define

$$\lambda(E) := \inf \left\{ \sum_{i=1}^{\infty} \operatorname{diam}(A_i) \ : \ E \subset \bigcup_{i \in \mathbb{N}} A_i \right\}.$$

It is easy to see that if $N = 2$ and if $E := \left\{ (t, \sin \frac{1}{t}) : t \in (0, 1] \right\}$, then $\lambda(E) < \infty$, while the length of $E$ is infinite. This is due to the fact that in the definition of $\lambda(E)$ the sets $A_i$ are not forced to follow the geometry of $E$. In view of this, $H^k(E)$ will be defined as a limit of outer measures $H^k_\delta(E)$ which will follow the geometry of $E$.

**Definition 4.21** *Let $E \subset \mathbb{R}^N$, $0 \le \delta < \infty$. We define*

$$H^s_\delta(E) := \inf \left\{ \sum_{i=1}^{\infty} \alpha(s) \left( \frac{\operatorname{diam} C_i}{2} \right)^s \ : \ E \subset \bigcup_{i \in \mathbb{N}} C_i, \operatorname{diam} C_i \le \delta \right\},$$

*where $\alpha(s) := \dfrac{\pi^{\frac{s}{2}}}{\Gamma(\frac{s}{2}+1)}$ and $\Gamma(s) := \int_0^\infty e^{-x} x^{s-1} \, dx \ (0 < s < \infty)$ is the usual Gamma function.*

**Remark 4.22** *If $\delta_1 \le \delta_2$, then $H^s_{\delta_2}(E) \le H^s_{\delta_1}(E)$ and so $\lim_{\delta \to 0} H^s_\delta(E) = \sup_{\delta > 0} H^s_\delta(E)$ exists.*

**Definition 4.23** *Let $E \subset \mathbb{R}^N$ and $0 \le s < \infty$. We define the $s$-dimensional Hausdorff measure on $\mathbb{R}^N$ by*

$$H^s(E) = \lim_{\delta \to 0} H^s_\delta(E).$$

**Theorem 4.24** *For each $0 \le s < \infty$, $H^s$ is a Borel measure.*

The proof of this result can be found in Evans and Gariepy (1992).

**Theorem 4.25**

   (i) *$H^0$ is the counting measure.*

   (ii) *$H^1 = H^1_\delta = \mathcal{L}^1$ on $\mathbb{R}$ for every $\delta > 0$.*

   (iii) *$H^s(E) = 0$ for all $E \subset \mathbb{R}^N$ and for all $s > N$.*

   (iv) *$H^s(\lambda E) = \lambda^s H^s(E)$ for all $E \subset \mathbb{R}^N$ and for all $\lambda > 0$.*

   (v) *$H^s(L(E)) = H^s(E)$ for all $E \subset \mathbb{R}^N$ and for all affine isometry $L : \mathbb{R}^N \to \mathbb{R}^N$.*

**Proof**

(i) We observe that $\alpha(0) = 1$. Therefore, $H^0(\{a\}) = 1$ for every $a \in \mathbb{R}^N$ and

$$H_\delta^0(\{a_1, \ldots, a_k\}) \geq k$$

for every mutually distinct $a_1, \ldots, a_k \in \mathbb{R}^N$ and for every positive $\delta$ such that $0 < \delta < \frac{1}{3} \min\{|a_i - a_j| : i \neq j\}$. Hence,

$$H^0(\{a_1, \ldots, a_k\}) = k,$$

i.e. $H^0$ is the counting measure.

(ii) Fix $\delta > 0$ and $E \subset \mathbb{R}$. We have

$$\mathcal{L}^1(E) = \inf \left\{ \sum_{i=1}^{\infty} |a_i - b_i| : E \subset \bigcup_{i \in \mathbb{N}} [a_i, b_i] \right\}$$

$$= \inf \left\{ \sum_{i=1}^{\infty} \operatorname{diam}(C_i) : E \subset \bigcup_{i \in \mathbb{N}} C_i \right\}$$

$$\leq \inf \left\{ \sum_{i=1}^{\infty} \operatorname{diam}(C_i) : E \subset \bigcup_{i \in \mathbb{N}} C_i, \operatorname{diam} C_i \leq \delta \right\}$$

$$= H_\delta^1(E).$$

On the other hand, setting $I_k := [\delta k, \delta(k+1)]$, $k \in \mathbb{Z}$, for every $C \subset \mathbb{R}$ we observe that $\operatorname{diam}(C \cap I_k) \leq \delta$ and

$$\sum_{k=-\infty}^{+\infty} \operatorname{diam}(C \cap I_k) \leq \operatorname{diam}(C).$$

Therefore, we have

$$\mathcal{L}^1(E) = \inf \left\{ \sum_{i=1}^{\infty} \operatorname{diam}(C_i) : E \subset \bigcup_{i \in \mathbb{N}} C_i \right\}$$

$$\geq \inf \left\{ \sum_{i=1}^{\infty} \sum_{k=-\infty}^{\infty} \operatorname{diam}(C_i \cap I_k) : E \subset \bigcup_{i \in \mathbb{N}} C_i \right\}$$

$$\geq H_\delta^1(E)$$

and this concludes (ii).

(iii) Let $Q = (0, 1)^N$ and let $m > 1$ be an integer. For $k = (k_1, \ldots, k_N) \in K := \{0, \ldots, m-1\}^N$ set

$$Q_k := \left[ \frac{k_1}{m}, \frac{k_1 + 1}{m} \right] \times \ldots \times \left[ \frac{k_N}{m}, \frac{k_N + 1}{m} \right].$$

We observe that

$$Q = \bigcup_{k \in K} Q_k \quad \text{and} \quad \text{diam}(Q_k) = \frac{\sqrt{N}}{m}$$

and so

$$H^s_{\frac{\sqrt{N}}{m}}(Q) \leq \sum_{k \in K} \alpha(s) \left(\frac{\sqrt{N}}{2m}\right)^s = \alpha(s) m^{N-s} \left(\frac{\sqrt{N}}{2}\right)^s.$$

Letting $m \to +\infty$ we deduce that

$$H^s(Q) = 0,$$

which yields

$$H^s(\mathbb{R}^N) = 0.$$

(iv) Using the property that $\text{diam}(\lambda C) = \lambda \, \text{diam}(C)$, it is easy to verify that $\lambda^s H^s(C) = H^s(\lambda C)$ for every $\lambda > 0$ and for every $C \subset \mathbb{R}^N$.

(v) This follows from the fact that $\text{diam}(L(C)) = \text{diam}(C)$ for every affine isometry $L : \mathbb{R}^N \to \mathbb{R}^N$ and for every $C \subset \mathbb{R}^N$.

$\square$

**Remark 4.26** It turns out that the equality

$$H^N(E) = \mathcal{L}^N(E)$$

holds for every $E \subset \mathbb{R}^N$. The proof is not trivial and it uses the *isodiametric inequality*

$$\mathcal{L}^N(E) \leq \alpha(N) \left(\frac{\text{diam } E}{2}\right)^N,$$

which is valid for every $E \subset \mathbb{R}^N$. We remark that $E$ does not have to be contained in a ball of diameter diam $E$.

**Lemma 4.27** *Let $E \subset \mathbb{R}^N$, $0 < \delta \leq \infty$, and let $0 \leq s < \infty$ be such that $H^s_\delta(E) = 0$. Then*

$$H^s(E) = 0.$$

**Proof** If $s = 0$, $H^0_\delta(E) = 0$ implies that $E = \emptyset$ and so $H^0(E) = 0$. Now assume that $s > 0$ and fix $\epsilon > 0$. There exist sets $\{C_i\}_{i \in \mathbb{N}}$ such that $E \subset \bigcup_{i \in \mathbb{N}} C_i$, diam $C_i \leq \delta$ for every $i \in \mathbb{N}$ and

$$\sum_{i=1}^{\infty} \alpha(s) \left(\frac{\text{diam } C_i}{2}\right)^s \leq \epsilon.$$

We observe that $\alpha(s)\left(\frac{\operatorname{diam} C_i}{2}\right)^s \leq \epsilon$ for every $i = 1,\ldots,\infty$ and so

$$\operatorname{diam} C_i \leq 2\left(\frac{\epsilon}{\alpha(s)}\right)^{\frac{1}{s}} =: l(\epsilon).$$

Hence,

$$H^s_{l(\epsilon)}(E) \leq \epsilon$$

and letting $\epsilon$ go to zero we deduce that $H^s(E) = 0$. □

**Lemma 4.28** *Let $E \subset \mathbb{R}^N$ and let $0 \leq s < t < \infty$ be two real numbers.*

　(i) *If $H^s(E) < +\infty$, then $H^t(E) = 0$.*
　(ii) *If $H^t(E) > 0$, then $H^s(E) = +\infty$.*

**Proof**　Assume that $H^s(E) < +\infty$ and fix $\delta > 0$. There exist sets $\{C_i\}_{i \in \mathbb{N}}$ such that $E \subset \cup_{i \in \mathbb{N}} C_i$, $\operatorname{diam} C_i < \delta$ and

$$\sum_{i=1}^{\infty} \alpha(s)\left(\frac{\operatorname{diam} C_i}{2}\right)^s \leq H^s_\delta(E) + 1.$$

Since

$$\begin{aligned}
H^t_\delta(E) &\leq \sum_{i=1}^{\infty} \alpha(t)\left(\frac{\operatorname{diam} C_i}{2}\right)^t \\
&= \frac{\alpha(t)}{\alpha(s)} 2^{s-t} \sum_{i=1}^{\infty} \alpha(s)\left(\frac{\operatorname{diam} C_i}{2}\right)^s (\operatorname{diam} C_i)^{t-s} \\
&\leq \frac{\alpha(t)}{\alpha(s)} 2^{s-t} \delta^{t-s} (H^s_\delta(E) + 1),
\end{aligned}$$

letting $\delta$ go to zero we obtain $H^t(E) = 0$ and assertion (ii) follows. □

**Definition 4.29** *Let $E \subset \mathbb{R}^N$. We define the Hausdorff dimension of $E$ by*

$$H_{dim}(E) := \inf\{0 \leq s < +\infty : H^s(E) = 0\}.$$

**Remark 4.30** Let $E \subset \mathbb{R}^N$.

　(i) By Theorem 4.25 (iii), $H_{dim}(E) \leq N$.
　(ii) If $s = H_{dim}(E)$, then $H^t(E) = 0$ for every $t > s$ while $H^t(E) = +\infty$ for every $t < s$. The dimension $H_{dim}(E)$ is not necessarily an integer and may be any number in $[0, N]$.

**Proposition 4.31** *Let $f : \mathbb{R}^N \to \mathbb{R}^M$ be a Lipschitz function, $E \subset \mathbb{R}^N$, and let $0 \leq s < \infty$. Then*

$$H^s(f(E)) \leq K^s H^s(E),$$

*where $K := \sup\left\{\frac{|f(x)-f(y)|}{|x-y|} : x \neq y, x, y \in \mathbb{R}^N\right\}$.*

**Proof** Without loss of generality, assume that $H^s(E) < +\infty$ and fix $\delta, \epsilon > 0$. There exist sets $\{C_i\}_{i\in\mathbb{N}}$ such that $E \subset \cup_{i\in\mathbb{N}}C_i$, diam $C_i \leq \delta$ for every $i \in \mathbb{N}$, and

$$\sum_{i=1}^{\infty} \alpha(s) \left( \frac{\operatorname{diam} C_i}{2} \right)^s \leq H^s_\delta(E) + \epsilon.$$

As $f$ is a Lipschitz function, diam $(f(C)) \leq K$ diam $(C)$ for every $C \subset \mathbb{R}^N$ and so

$$H^s_{K\delta}(f(E)) \leq \sum_{i=1}^{\infty} \alpha(s) \left( \frac{\operatorname{diam} f(C_i)}{2} \right)^s \leq K^s(H^s_\delta(E) + \epsilon).$$

Letting $\epsilon$ and $\delta$ go to zero we conclude that $H^s(f(E)) \leq K^s H^s(E)$.          □

Setting $f = \chi_E$ in the Lebesgue–Besicovitch Differentiation Theorem (see Theorem 4.18), it follows immediately that

**Proposition 4.32** *If $E \subset \mathbb{R}^N$ is a measurable set, then*

$$\lim_{\epsilon \to 0^+} \frac{\mathcal{L}^N(B(x_0, \epsilon) \cap E)}{\mathcal{L}^N(B(x_0, \epsilon))} = \begin{cases} 1 & \text{for } \mathcal{L}^N \text{a.e. } x_0 \in E \\ 0 & \text{for } \mathcal{L}^N \text{a.e. } x_0 \in \mathbb{R}^N \backslash E. \end{cases}$$

As it turns out, this result can be extended to lower dimensional Hausdorff measures.

**Proposition 4.33** *Let $E \subset \mathbb{R}^N$ be an $H^s$-measurable set, $0 \leq s < N, H^s(E) < +\infty$. Then*

$$\lim_{\epsilon \to 0^+} \frac{H^s(B(x_0, \epsilon) \cap E)}{\alpha(s)\epsilon^s} = 0 \tag{4.2}$$

*for $H^s$a.e. $x_0 \in \mathbb{R}^N \backslash E$ and for $H^s$ a.e. $x_0 \in E$*

$$\limsup_{\epsilon \to 0^+} \frac{H^s(B(x_0, \epsilon) \cap E)}{\alpha(s)\epsilon^s} \leq 1. \tag{4.3}$$

To prove this proposition we need the following covering theorem.

**Theorem 4.34 [Vitali's Covering Theorem]** *Let $\mathcal{F}$ be a collection of closed balls in $\mathbb{R}^N$ with $\sup\{\operatorname{diam} B : B \in \mathcal{F}\} < +\infty$. Then there exists a countable family $\mathcal{G}$ of disjoint balls in $\mathcal{F}$ such that*

$$\bigcup_{B\in\mathcal{F}} B \subset \bigcup_{B\in\mathcal{G}} \hat{B},$$

*where, if $B = B(x, r)$, $\hat{B}$ denotes $B(x, 5r)$.*

**Proof** Let $D := \sup\{\text{diam } B : B \in \mathcal{F}\}$ and set

$$\mathcal{F}_j := \left\{ B \in \mathcal{F} : \frac{D}{2^j} < \text{diam } B \leq \frac{D}{2^{j-1}} \right\}.$$

We define $\mathcal{G}$ as follows. Let $\mathcal{G}_1$ be a maximal disjoint collection of balls in $\mathcal{F}_1$ and, assuming that $\mathcal{G}_1,...,\mathcal{G}_{k-1}$ have been chosen, we select $\mathcal{G}_k$ to be any maximal disjoint collection of

$$\left\{ B \in \mathcal{F}_k : \ B \cap B' = \emptyset \text{ for all } B' \in \bigcup_{j=1}^{k-1} \mathcal{G}_j \right\}.$$

We set $\mathcal{G} := \cup_{i \in \mathbb{N}}\mathcal{G}_j$. It is clear that $\mathcal{G}$ is a collection of disjoint balls of $\mathcal{F}$ and we claim that, given $B \in \mathcal{F}$, there exists $B' \in \mathcal{G}$ such that $B \cap B' \neq \phi$ and $B \subset \hat{B}'$. Indeed, let $k$ be such that $B \in \mathcal{F}_k$ and $B \notin \mathcal{G}_k$. Due to the maximality of $\mathcal{G}_k$, there exists a ball $B' \in \cup_{j=1}^{k-1}\mathcal{G}_k$ such that $B \cap B' \neq \phi$. Hence,

$$\text{diam } B' \geq \frac{D}{2^k}, \quad \text{diam } B \leq \frac{D}{2^{k-1}}$$

and we conclude that $\text{diam } B \leq 2 \text{ diam } B'$. Thus $B \subset \hat{B}'$ and $\mathcal{G}$ is a countable family since any collection of disjoint open sets must be countable. □

**Corollary 4.35** *Let $\mathcal{F}$ be a fine cover of $A$ by closed balls, i.e.*

$$\inf\{\text{diam } B : \ B \in \mathcal{F}, x \in B\} = 0 \text{ for every } x \in A,$$

*and suppose, further, that $\sup\{\text{diam } B : B \in \mathcal{F}\} < +\infty$. There exists a countable family $\mathcal{G}$ of disjoint balls in $\mathcal{F}$ such that for each finite subset $\{B_1, \dots, B_n\} \subset \mathcal{F}$ we have*

$$A \backslash (\bigcup_{i=1}^{n} B_i) \subset \bigcup_{B \in \mathcal{G} \backslash \{B_1,\dots,B_n\}} \hat{B}.$$

**Proof** By Vitali's Covering Theorem there exists a countable family $\mathcal{G}$ of disjoint balls in $\mathcal{F}$ such that

$$A \subset \bigcup_{B \in \mathcal{F}} B \subset \bigcup_{B \in \mathcal{G}} \hat{B}.$$

Fixing $B_1, \dots, B_n \in \mathcal{F}$, if $x \notin A \backslash (\bigcup_{i=1}^{n} B_i)$, then, as $B_i$ are closed and $\mathcal{F}$ is a fine cover of $A$, there exists $B \in \mathcal{F}$ with $x \in B$, $B \cap B_i = \emptyset$ for $i = 1, \dots, n$. From the claim in the proof of Vitali's Theorem, it follows that there exists $B^* \in \mathcal{G}$ with $B^* \cap B \neq \emptyset$; hence $B^* \neq B_1, \dots, B_n$ and $B \subset \hat{B}^*$. □

**Proof of Proposition 4.33.** It is clear that to prove (4.2) it suffices to show that

$$H^s(A_t) = 0$$

for every $t > 0$, where

$$A_t := \left\{ x \in \mathbb{R}^N \backslash E : \limsup_{r \to 0^+} \frac{H^s(B(x,r) \cap E)}{r^s} > t \right\}.$$

Fix $\epsilon > 0$. As $H^s \lfloor E$ is a Radon measure, we may find a compact set $K \subset E$ such that $H^s(E \backslash K) \leq \epsilon$. Thus $A_t \subset (\mathbb{R}^N \backslash K)$ and, for fixed $\delta > 0$, we consider

$$\mathcal{F} := \left\{ \bar{B}(x,r) : \bar{B}(x,r) \subset (\mathbb{R}^N \backslash K), 0 < r < \delta, \frac{H^s(B(x,r) \cap E)}{\alpha(s)r^s} > t \right\}.$$

Without loss of generality, we may assume that $\mathcal{F} \neq 0$; otherwise $A_t = \phi$ and so $H^s(A_t) = 0$.

Using Vitali's Covering Theorem we write

$$A_t \subset \bigcup_{i \in \mathbb{N}} \hat{B}_i$$

where $\{B(x_i, r_i)\}_{i=1}^\infty \subset \mathcal{F}$ is a disjoint family of closed balls. Then

$$H_{10\delta}^s(A_t) \leq \sum_{i=1}^\infty \alpha(s)(5r_i)^s$$

$$\leq \frac{5^s}{t} \sum_{i=1}^\infty H^s(B_i \cap E)$$

$$\leq \frac{5^s}{t} H^s((\mathbb{R}^N \backslash K) \cap E)$$

$$= \frac{5^s}{t} H^s(E \backslash K)$$

$$\leq \frac{5^s}{t} \epsilon.$$

Letting $\delta \to 0^+$ and then $\epsilon \to 0^+$ we conclude that $H^s(A_t) = 0$. In order to prove the inequality in (4.3), namely,

$$\limsup_{\epsilon \to 0^+} \frac{H^s(B(x_0, r) \cap E)}{\alpha(s)r^s} \leq 1$$

for $H^s$ a.e. $x_0 \in E$, we proceed as in the first part of this proof. We set

$$B_t := \left\{ x \in E : \limsup_{r \to 0} \frac{H^s(B(x,r) \cap E)}{\alpha(s)r^s} > t \right\}.$$

Since $H^s \lfloor E$ is a Radon measure, there exists an open set $U$ containing $B_t$ such that

$$H^s(U \cap E) \leq H^s(B_t) + \epsilon$$

and we define, for fixed $\delta > 0$,

$$\mathcal{F} := \left\{ B(x,r) : \ B(x,r) \subset U, 0 < r < \delta, \ \frac{H^s(B(x,r) \cap E)}{\alpha(s)r^s} > t \right\}.$$

By Corollary 4.35, for each $m \in \mathbb{N}$ there exists a countable disjoint family of balls $\{B_i\}_{i=1}^{\infty}$ in $\mathcal{F}$ such that

$$B_t \subset \bigcup_{i=1}^{m} B_i \ \bigcup_{i=m+1}^{\infty} \hat{B}_i \ ,$$

where $B_i = B(x_i, r_i)$. Then

$$H^s_{10\delta}(B_t) \leq \sum_{i=1}^{m} \alpha(s)r_i^s + \sum_{i=m+1}^{\infty} \alpha(s)(5r_i)^s$$

$$\leq \frac{1}{t}\sum_{i=1}^{m} H^s(B_i \cap E) + \frac{5^s}{t} \sum_{i=m+1}^{\infty} H^s(B_i \cap E)$$

$$\leq \frac{1}{t}H^s(U \cap E) + \frac{5^s}{t}H^s \left( \bigcup_{i=m+1}^{\infty} B_i \cap E \right).$$

Letting $m \to +\infty$ we deduce that

$$H^s_{10\delta}(B_t) \leq \frac{1}{t}H^s(U \cap E)$$

$$\leq \frac{1}{t}(H^s(B_t) + \epsilon).$$

Letting $\delta \to 0^+$ and then $\epsilon \to 0^+$, we conclude that

$$H^s(B_t) \leq \frac{1}{t}H^s(B_t);$$

hence, as $H^s(B_t) \leq H^s(E) < +\infty$,

$$H^s(B_t) = 0 \text{ if } t > 1.$$

$\square$

**Remark 4.36** Under the hypotheses of Proposition 4.33, it can be shown that

$$\limsup_{\epsilon \to 0^+} \frac{H^s(B(x_0, \epsilon) \cap E)}{\alpha(s)\epsilon^s} \geq \frac{1}{2^s}$$

for $H^s$ a.e. $x_0 \in E$. It is worth noting that it is possible to have

$$\limsup_{\epsilon \to 0+} \frac{H^s(B(x_0, \epsilon) \cap E)}{\alpha(s)\epsilon^s} < 1$$

and

$$\liminf_{\epsilon \to 0+} \frac{H^s(B(x_0, \epsilon) \cap E)}{\alpha(s)\epsilon^s} = 0$$

for $H^s$ a.e. $x_0 \in E$, where $0 < H^s(E) < +\infty$.

In the last result of this section we estimate the Hausdorff measure of a set where a locally integrable function is concentrated.

**Proposition 4.37** *Let* $f \in L^1_{loc}(\mathbb{R}^N)$, $0 \le s < N$ *and set*

$$\Lambda_s := \left\{ x \in \mathbb{R}^N : \limsup_{r \to 0+} \frac{1}{r^s} \int_{B(x,r)} |f(y)| \, dy > 0 \right\}.$$

*Then* $H^s(\Lambda_s) = 0$.

**Proof** Without loss of generality, we may assume that $f \in L^1(\mathbb{R}^N)$, as it suffices to prove the result for $f \cdot \psi_k$, $\psi_k \in C_c(\mathbb{R}^N)$, $0 \le \psi_k \le 1$, $\psi_k(x) = 1$, if $x \in B(0, k)$.

By the Lebesgue–Besicovitch Differentiation Theorem (see Theorem 4.18),

$$\lim_{r \to 0} \frac{1}{\mathcal{L}^N(B(x,r))} \int_{B(x,r)} |f(y)| \, dy = |f(x)|$$

for $\mathcal{L}^N$ a.e. $x \in \mathbb{R}^N$ and so

$$\lim_{r \to 0} \frac{1}{r^s} \int_{B(x,r)} |f(y)| \, dy = 0.$$

Thus $\mathcal{L}^N(\Lambda_s) = 0$. Fix $\epsilon, \delta, \sigma > 0$, and as $f \in L^1(\mathbb{R}^N)$ choose $\eta > 0$ such that

$$\int_U |f(x)| \, dx < \sigma$$

whenever $\mathcal{L}^N(U) \le \eta$. Define

$$\Lambda_s^\epsilon := \left\{ x \in \mathbb{R}^N : \limsup_{r \to 0} \frac{1}{r^s} \int_{B(x,r)} |f(y)| \, dy > \epsilon \right\}.$$

Since $\Lambda_s^\epsilon \subset \Lambda_s$ we have

$$\mathcal{L}^N(\Lambda_s^\epsilon) = 0$$

and so we may find an open set $U \supset \Lambda_s^\epsilon$ such that $\mathcal{L}^N(U) < \eta$. We set

$$\mathcal{F} := \left\{ \bar{B}(x,r) : x \in \Lambda_s^\epsilon, 0 < r < \delta, \bar{B}(x,r) \subset U, \int_{B(x,r)} |f(y)| \, dy > \epsilon r^s \right\}.$$

By Vitali's Covering Theorem (see Theorem 4.34),

$$\Lambda_s^\epsilon \subset \bigcup_{i \in \mathbb{N}} \hat{B}_i, \quad B_i = B(x_i, r_i),$$

where $\{B_i\}_{i=1}^\infty \subset \mathcal{F}$ is a countable family of disjoint closed balls, and so

$$H_{10\delta}^s(\Lambda_s^\epsilon) \leq \sum_{r=1}^\infty \alpha(s)(5r_i)^s$$

$$\leq \frac{\alpha(s)5^s}{\epsilon} \sum_{i=1}^\infty \int_{B_i} |f(y)| \, dy$$

$$\leq \frac{\alpha(s)5^s}{\epsilon} \int_U |f(y)| \, dy$$

$$\leq \frac{\alpha(s)5^s}{\epsilon} \sigma.$$

It suffices to let $\delta \to 0$ and then $\sigma \to 0$ to obtain $H^s(\Lambda_s^\epsilon) = 0$. □

## 4.3 Overview of Sobolev spaces

Throughout this section $D \subset \mathbb{R}^N$ is an open set, $\theta$ is a symmetric mollifier on $\mathbb{R}^N$ (see Definition 1.16), and $\theta_r : \mathbb{R}^N \to \mathbb{R}$ is defined by

$$\theta_r(x) := \frac{1}{r^N} \theta\left(\frac{x}{r}\right), \quad x \in \mathbb{R}^N, r > 0.$$

Let

$$D_r := \{x \in D : \text{dist}(x, \partial D) > r\}$$

and for every $f \in L^1_{loc}(D)$.

$$f_r(x) := f * \theta_r(x) = \int_{D_r} \theta_r(x - y) f(y) \, dy.$$

**Definition 4.38** *Let $1 \leq p \leq +\infty$ and let $D \subset \mathbb{R}^N$ be an open set. A function $f \in L^1_{loc}(D)$ belongs to the Sobolev space $W^{1,p}(D)$ if $f \in L^p(D)$ and the distributional partial derivatives $\frac{\partial f}{\partial x_i}$ belong to $L^p(D), i = 1, \ldots, N$.*

We recall that when $f \in L^1_{loc}(D)$, $i \in \{1, \ldots, N\}$, then $\frac{\partial f}{\partial x_i}$ is the distribution in $\mathcal{D}'(D)$ defined by

$$< \frac{\partial f}{\partial x_i}, \varphi >:= -\int_D f(x) \frac{\partial \varphi}{\partial x_i}(x) \, dx$$

for every $\varphi \in C_c^\infty(D)$.

**Theorem 4.39** *Let $f \in L^1_{loc}(D)$.*

(i) *For each $r > 0$, $f_r \in C^\infty(D_r)$.*

(ii) *If $f \in C(D)$, then $f_r$ converges to $f$ uniformly on compact subsets of $D$, when $r$ tends to zero.*

(iii) *If $1 \le p < +\infty$ and if $f \in L^p_{loc}(D)$, then $f_r$ converges to $f$ in $L^p_{loc}(D)$.*

(iv) *If $1 \le p < +\infty$, $f \in L^p_{loc}(D)$ and if $x \in D$ is a Lebesgue point of $f$, then*

$$\lim_{r \to 0} f_r(x) = f(x).$$

(v) *If $1 \le p \le +\infty$ and if $f \in W^{1,p}_{loc}(D)$, then*

$$\frac{\partial f_r}{\partial x_i} = \frac{\partial f}{\partial x_i} * \theta_r \quad (i = 1, \dots, N)$$

*in $D_r$.*

(vi) *If $1 \le p < +\infty$ and if $f \in W^{1,p}_{loc}(D)$, then $f_r$ converges to $f$ in $W^{1,p}_{loc}(D)$.*

For the proof of this result we refer the reader to Adams (1975).

**Definition 4.40** *We say that $\partial D$ is Lipschitz, or $D$ is a Lipschitz domain, if for each point $x \in \partial D$ there exists a Lipschitz mapping $\gamma : \mathbb{R}^{N-1} \to \mathbb{R}$ such that, upon rotating and relabelling of the coordinate axes if necessary, we have*

$$D \cap Q(x,r) = \{y \in \mathbb{R}^N : \gamma(y_1, \dots, y_{N-1}) < y_N\} \cap Q(x,r).$$

**Theorem 4.41** *Let $1 \le p < +\infty$ and let $f \in W^{1,p}(D)$, where $D$ is bounded and Lipschitz. Then there exists a sequence $\{f_k\}_{k \in \mathbb{N}} \subset W^{1,p}(D) \cap C^\infty(\bar{D})$ such that $f_k \to f$ in $W^{1,p}(D)$.*

This theorem is proved in Adams (1975).

**Theorem 4.42** *Let $D \subset \mathbb{R}^N$ be an open set and assume that $1 \le p < \infty$.*

(i) *If $f, g \in W^{1,p}(D) \cap L^\infty(D)$, then $fg \in W^{1,p}(D) \cap L^\infty(D)$ and $\frac{\partial(fg)}{\partial x_i} = \frac{\partial f}{\partial x_i}g + f\frac{\partial g}{\partial x_i}$ for $\mathcal{L}^N$ a.e. $x \in \Omega, i = 1, \dots, N$.*

(ii) *If $f \in W^{1,p}(D)$, $F \in C^1(\mathbb{R})$, $F' \in L^\infty(\mathbb{R})$ and $F(0) = 0$, then $F \circ f \in W^{1,p}(D)$ and $\frac{\partial F \circ f}{\partial x_i} = F' \circ \frac{\partial f}{\partial x_i}$ for $\mathcal{L}^N$ a.e. $x \in D, i = 1, \dots, N$. If, in addition, $\mathcal{L}^N(D) < \infty$, then the condition $F(0) = 0$ is unnecessary.*

(iii) *If $f \in W^{1,p}(D)$, then $f^+, f^-, |f| \in W^{1,p}(D)$,*

$$\nabla f^+(x) = \begin{cases} \nabla f(x) & \mathcal{L}^N \text{ a.e. } x \in \{f > 0\} \\ 0 & \mathcal{L}^N \text{ a.e. } x \in \{f \le 0\}, \end{cases}$$

$$\nabla f^-(x) = \begin{cases} 0 & \mathcal{L}^N \text{ a.e. } x \in \{f \ge 0\} \\ \nabla f & \mathcal{L}^N \text{ a.e. } x \in \{f < 0\}, \end{cases}$$

$$\nabla \left| f \right| (x) = \begin{cases} \nabla f & \mathcal{L}^N \text{ a.e. } x \in \{f > 0\} \\ 0 & \mathcal{L}^N \text{ a.e. } x \in \{f = 0\} \\ -\nabla f & \mathcal{L}^N \text{ a.e. } x \in \{f < 0\} \end{cases}$$

*and*

$$\nabla f(x) = 0, \ \mathcal{L}^N \text{ a.e. } x \in \{f = 0\}.$$

**Proof of part (iii)** As $f^- = (-f)^+$ and $\left| f \right| = f^+ + f^-$, it suffices to prove that $f^+ \in W^{1,p}(D)$ and

$$\nabla f^+(x) = \begin{cases} \nabla f & \mathcal{L}^N \text{a.e. } x \in \{f \geq 0\} \\ 0 & \mathcal{L}^N \text{a.e. } x \in \{f \leq 0\}. \end{cases}$$

Fix $\epsilon > 0$ and define

$$F_\epsilon(r) := \begin{cases} \sqrt{r^2 + \epsilon^2} - \epsilon & r \geq 0 \\ 0 & r < 0. \end{cases}$$

By (ii) and given $\varphi \in C_0^1(\Omega)$, we have

$$\int_\Omega F_\epsilon(f(x)) \frac{\partial \varphi}{\partial x_i}(x) \, dx = - \int_\Omega F_\epsilon'(f(x)) \frac{\partial f}{\partial x_i}(x) \varphi(x) \, dx$$

for every $i \in \{1, \ldots, N\}$, and letting $\epsilon \to 0^+$ we conclude that

$$\frac{\partial f^+}{\partial x_i} = \frac{\partial f}{\partial x_i} \text{ for } \mathcal{L}^N \text{ a.e. } x \in \{z : f(z) \geq 0\}$$

and $\frac{\partial f^+}{\partial x_i} = 0$ otherwise. Thus $f^+ \in W^{1,p}(D)$.                    $\square$

**Proposition 4.43** *Let $D \subset \mathbb{R}^N$ be an open set, $1 \leq p < +\infty$.*
  (i) *If $f, g \in W^{1,p}(D)$, then $H := \max\{f, g\}, h := \min\{f, g\} \in W^{1,p}(D)$, and*

$$\nabla H = \begin{cases} \nabla f & \mathcal{L}^N \text{ a.e. } x \in \{f \geq g\} \\ \nabla g & \mathcal{L}^N \text{ a.e. } x \in \{f \leq g\}, \end{cases}$$

$$\nabla h = \begin{cases} \nabla g & \mathcal{L}^N \text{ a.e. } x \in \{f \geq g\} \\ \nabla f & \mathcal{L}^N \text{ a.e. } x \in \{f \leq g\}. \end{cases}$$

  (ii) *If $f_k \in W^{1,p}(D)$, then $h := \sup_k \left| \nabla f_k \right| \in L^p(D)$ and if $g := \sup_k f_k \in L^p(D)$, then $g \in W^{1,p}(D)$ and $\left| \nabla g(x) \right| \leq h(x)$ for $\mathcal{L}^N$ a.e. $x \in D$.*

## Proof

(i) As $H = \max\{f, b\} = f + (g - f)^+$, it suffices to apply Theorem 4.42 (iii) to $H$, and, similarly, to $h$.

(ii) Let $g_m := \max\{f_k : 1 \le h \le m\}$. By part (i) $g_m \in W^{1,p}(D)$ and

$$|\nabla g_m(x)| \le \max\{|\nabla f_k(x)| : 1 \le k \le m\}$$

$$\le h(x).$$

Since $g_m \to g$ pointwise and increasing, and $\{|\nabla g_m|\}$ is a sequence bounded in $L^p(D)$, we conclude that

$$g_m \to g \text{ in } L^p(D) \,, \nabla g_m \rightharpoonup \nabla g \text{ in } L^p(D)$$

and so

$$|\nabla g(x)| \le h(x) \text{ a.e. in } D.$$

$\square$

**Theorem 4.44** [**Sobolev–Niremberg–Gagliardo Inequality**] *If* $1 \le p < N$, *then there exists a constant* $C = C(N, p)$ *such that*

$$\left\{ \int_{\mathbb{R}^N} |f(x)|^{p^*} \, dx \right\}^{1/p*} \le C \left\{ \int_{\mathbb{R}^N} |\nabla f(x)|^p \, dx \right\}^{1/p}$$

*for every* $f \in W^{1,p}(\mathbb{R}^N)$, *where* $p^* := \frac{Np}{N-p}$ *is the Sobolev exponent.*

This result is part of a series of imbedding theorems for Sobolev functions, of which we select the following

**Theorem 4.45** *Let* $\Omega \subset \mathbb{R}^N$ *be a bounded, Lipschitz domain.*

(i) $W^{1,p}(\Omega) \subset L^{p^*}(\Omega)$, *with* $\frac{1}{p^*} = \frac{1}{p} - \frac{1}{N}$, *if* $1 \le p < N$.

(ii) $W^{1,p}(\Omega) \subset L^q(\Omega)$, *for all* $q \in [p, +\infty)$, *if* $p = N$.

(iii) $W^{1,p}(\Omega) \subset L^\infty(\Omega)$, *if* $p > N$, *and the imbeddings in* (i)–(iii) *are continuous. Moreover, the imbedding in* (ii) *is compact.*

(iv) $W^{1,p}(\Omega) \subset L^q(\Omega)$ *with a compact imbedding if* $q < p^*, 1 \le p < N$.

(v) $W^{1,p}(\Omega) \subset C(\bar\Omega)$ *with a compact imbedding, if* $p > N$.

The proof of Theorem 4.44 can be found in Evans and Gariepy (1992), while the latter and the following extension theorem are proven in Adams (1975).

**Remark 4.46** We note that (i)–(iii) also hold for $\Omega = \mathbb{R}^N$ and $W^{1,p}(\mathbb{R}^N) \subset C(\mathbb{R}^N)$ if $p > N$.

**Theorem 4.47 [Extension Theorem]** *Let $D \subset \mathbb{R}^N$ be a Lipschitz domain, $1 \leq p \leq +\infty$, and let $D \subset\subset V$, $V \subset \mathbb{R}^N$ be an open set. There exists a bounded, linear operator*

$$E : W^{1,p}(D) \to W^{1,p}(\mathbb{R}^N)$$

*such that*

$$E(f) = f \ \text{in} \ D, \ \text{spt} \, E(f) \subset V$$

*and*

$$\|E(f)\|_{W^{1,p}(\mathbb{R}^N)} \leq C(N, p, D) \, \|f\|_{W^{1,p}(D)}.$$

**Remark 4.48** If $f \geq 0$, then $E(f) \geq 0$. Also, the next result asserts that a Sobolev function $f \in W^{1,p}(\mathbb{R}^N)$ can be expanded in a finite Taylor series such that, for all points in the complement of a set of measure zero, the integral average of the remainder term tends to 0.

**Theorem 4.49** *Let $D \subset \mathbb{R}^N$ be an open set, $1 \leq p \leq +\infty$, and let $f \in W^{1,p}(D)$. Then for almost every $x \in D$ and for every $0 < |h| < \text{dist}(x, \partial D)$ we have*

(i) $R_{h,x} f \in W^{1,p}(B(0,1))$,

(ii) $\lim\limits_{h \to 0} \|R_{h,x} f\|_{1,p}(B(0,1)) = 0$, *where*

$$R_{h,x} f(X) := \frac{f(x + hX) - f(x)}{h} - \sum_{i=1}^{N} \frac{\partial f}{\partial x_i}(x) X_i$$

*for all $X \in B(0,1)$.*

For the proof, we refer the reader to Gold'sthein and Reshetnyak (1990) or Ziemer (1989).

**Theorem 4.50** *Assume that $D$ is bounded, $\partial D$ is Lipschitz, and $1 \leq p < +\infty$.*

(i) *There exists a bounded linear operator $T : W^{1,p}(D) \to L^p(\partial D; H^{N-1})$ such that*

$$T(f) = f \quad \text{on} \ \partial D$$

*for every $f \in W^{1,p}(D) \cap C(\bar{D})$.*

(ii) *Furthermore, for every $\phi \in C^1(\mathbb{R}^N)^N$ and every $f \in W^{1,p}(D)$ we have*

$$\int_D f \, \text{div}(\phi) \, dx = -\int_D \nabla f \cdot \phi \, dx + \int_{\partial D} (\phi \cdot \nu) T(f) \, dH^{N-1},$$

*where $\nu$ denotes the unit outer normal to $\partial D$.*

(iii) *If $f \in W^{1,p}(D)$, then*

$$\lim_{r \to 0^+} \frac{1}{\mathcal{L}^N(B(x,r))} \int_{B(x,r) \cap D} |T(f)(x) - f(y)|_2 \, dy = 0$$

*for $H^{N-1}$ a.e. $x \in \partial D$.*

This result is proved in Evans and Gariepy (1992).

**Definition 4.51** *The function $T(f)$, which is uniquely defined up to subsets of $\partial D$ of $H^{N-1}$ measure zero, is called the* trace *of $f$ on $\partial D$. We say that $T(f)$ is the* boundary value *of $f$ on $\partial D$.*

## 4.4   $p$-capacity

The notion of $p$-capacity, $1 \le p < N$, allows us to characterize the smallness of subsets of $\mathbb{R}^N$ and it is an important tool in the study of continuity properties of $W^{1,p}$ functions.

**Definition 4.52** *Let $1 \le p < N$ and $p^* = \frac{Np}{N-p}$. We define*

$$K^p := \left\{ f : \mathbb{R}^N \to \mathbb{R} \ : \ f \ge 0, f \in L^{p^*}(\mathbb{R}^N), \nabla f \in L^p(\mathbb{R}^N)^N \right\}.$$

Note that since $L^{p^*}(\mathbb{R}^N) \not\subset L^p(\mathbb{R}^N)$, we cannot deduce from the Sobolev Imbedding Theorem that $K^p \subset W^{1,p}(\mathbb{R}^N)$.

**Definition 4.53** *Let $E \subset \mathbb{R}^N$ and $1 \le p < N$. We define the $p$-capacity of $E$ by*

$$Cap_p(E) := \inf \left\{ \int_{\mathbb{R}^N} |\nabla f|^p \, dx \ : \ f \in K^p, E \subset \text{int}\{f \ge 1\} \right\}.$$

**Remark 4.54**

(i) If $K \subset \mathbb{R}^N$ is a compact set, then, using mollifications of the characteristic function $\chi_k$, one has

$$Cap_p(K) = \inf \left\{ \int_{\mathbb{R}^N} |\nabla f|^p \, dx \ : \ f \in C_c^\infty(\mathbb{R}^N), f \ge \chi_K \right\}.$$

(ii) It is clear that if $E \subset F \subset \mathbb{R}^N$, then $Cap_p(E) \le Cap_p(F)$.

The following Sobolev–Niremberg–Gagliardo generalization will be used to prove that $Cap_p$ is an outer measure.

**Lemma 4.55** *Let $1 \le p < N$. Then there exists $C = C(N, p)$ such that*

$$\left\{ \int_{\mathbb{R}^N} |f(x)|^{p^*} \, dx \right\}^{1/p^*} \le C(N, p) \left\{ \int_{\mathbb{R}^N} |\nabla f(x)^p| \, dx \right\}^{1/p}$$

*for every $f \in K^p$.*

**Proof** We start by constructing a sequence $\varphi_n \in C_c^\infty(\mathbb{R}^N), 0 \le \varphi_n \le 1$, $\{\varphi_n(x)\}$ converges increasingly to 1 for a.e. $x \in \mathbb{R}^N$, and

$$\varphi_n(x) = 1 \text{ if } |x| < n, \quad \sup_n \int_{\mathbb{R}^N} |\nabla \varphi_n(x)|^N \, dx < +\infty.$$

One such sequence may be obtained as follows: let $\varphi \in C_c^\infty(B(0,2)), \varphi = 1$ in $B(0,1), 0 \le \varphi \le 1$ and set $\varphi_n(x) = \varphi\left(\frac{x}{n}\right)$. Since $p^* > p$, by Hölder's Inequality $\varphi_n f \in W^{1,p}(\mathbb{R}^N)$ for every $n$, and so, by Theorem 4.44,

$$\left\{\int_{\mathbb{R}^N} |f(x)\varphi_n(x)|^{p^*} \, dx\right\}^{1/p^*} \le C\left\{\int_{\mathbb{R}^N} |\nabla(f\varphi_n)(x)|^p \, dx\right\}^{1/p}$$

$$\le C\left\{\int_{\mathbb{R}^N} |\nabla f(x)|^p \, dx\right\}^{1/p}$$

$$+ C\left\{\int_{\mathbb{R}^N} |f(x)\nabla\varphi_n(x)|^p \, dx\right\}^{1/p}$$

and so, by Lebesgue's Dominated Convergence Theorem,

$$\left\{\int_{\mathbb{R}^N} |f(x)|^{p^*} \, dx\right\}^{1/p^*} \le C\left\{\int_{\mathbb{R}^N} |\nabla f(x)|^p \, dx\right\}^{1/p}$$

$$+ C \liminf_{n \to +\infty} \left\{\int_{\mathbb{R}^N} |f(x)\nabla\varphi_n(x)|^p \, dx\right\}^{1/p}.$$

It remains to show that

$$\int_{\mathbb{R}^N} |f(x)\nabla\varphi_n(x)|^p \, dx \underset{n \to +\infty}{\to} 0.$$

Indeed,

$$\int_{\mathbb{R}^N} |f(x)\nabla\varphi_n(x)|^p \, dx$$

$$\le \left\{\int_{\{|x|>n\}} |f(x)|^{p^*} \, dx\right\}^{p/p^*} \left\{\int_{\mathbb{R}^N} |\nabla\varphi_n(x)|^{p\left(\frac{p^*}{p}\right)'} \, dx\right\}^{1-p/p^*}$$

and $p\left(\frac{p^*}{p}\right)' = N$. Thus,

$$\int_{\mathbb{R}^N} |f(x)\nabla\varphi_n(x)|^p \, dx \le C'\left\{\int_{\{|x|>n\}} |f(x)|^{p^*} \, dx\right\}^{p/p^*}$$

and, due to the integrability of $|f|^{p^*}$, we conclude the result. $\square$

**Theorem 4.56** *Let* $1 \le p < N$. *Then* $Cap_p$ *is an outer measure on* $\mathbb{R}^N$.

**Proof** Let $\{E_k, k \in \mathbb{N}\}$ be a sequence of subsets of $\mathbb{R}^N$ and

$$E := \bigcup_{k \in \mathbb{N}} E_k.$$

We want to show that

$$Cap_p(E) \leq \sum_{k=1}^{\infty} Cap_p(E_k)$$

and so, we may assume, without loss of generality, that

$$\sum_{k \in \mathbb{N}} Cap_p(E_k) < +\infty.$$

Fix $\epsilon > 0$. For each $k \in \mathbb{N}$, we choose a function $f_k \in K^p$ such that $E_k \subset int\{f_k \geq 1\}$ and

$$\int_{\mathbb{R}^N} |\nabla f_k(x)|_2^p \, dx \leq Cap_p(E_k) + \frac{\epsilon}{2^k}.$$

Let $g := \sup_{k \in \mathbb{N}} f_k$. We observe that $E \subset int\{g \geq 1\}$ and, by Lemma 4.55,

$$\begin{aligned}
\int_{\mathbb{R}^N} g(x)^{p^*} \, dx &= \int_{\mathbb{R}^N} \sup_k f_k^{p^*}(x) \, dx \\
&\leq \sum_{k=1}^{\infty} \int_{\mathbb{R}^N} f_k^{p^*}(x) \, dx \\
&\leq c \sum_{k=1}^{\infty} \left\{ \int_{\mathbb{R}^N} |\nabla f_k(x)|^p \, dx \right\}^{p^*/p} \\
&\leq c \sum_{k=1}^{\infty} \left\{ Cap_p(E_k) + \frac{\epsilon}{2^k} \right\}^{p^*/p} \\
&\leq c \left\{ \sum_{k=1}^{\infty} \left[ Cap_p(E_k) + \frac{\epsilon}{2^k} \right] \right\}^{p^*/p} \\
&< +\infty.
\end{aligned}$$

Thus, if $\Omega' \subset\subset \Omega$, then $g \in L^p(\Omega')$ and setting $h := \sup |\nabla f_k|$ we have

$$\int_{\mathbb{R}^N} |h(x)|^p \, dx \leq \sum_{k=1}^{\infty} \int_{\mathbb{R}^N} |\nabla f_k(x)|^p \, dx$$

hence, by Proposition 4.43, we conclude that

$$g \in W_{loc}^{1,p}(\mathbb{R}^N), \quad |\nabla g(x)| \leq h(x) \text{ for } \mathcal{L}^N \text{a.e. } x \in \mathbb{R}^N.$$

Therefore, $g \in K^p$ and

$$Cap_p(E) \le \int_{\mathbb{R}^N} |\nabla g(x)|^p \, dx$$

$$\le \int_{\mathbb{R}^N} \sup |\nabla f_k(x)|^p \, dx$$

$$\le \sum_{k=1}^{\infty} \int_{\mathbb{R}^N} |\nabla f_k(x)|^p \, dx$$

$$\le \sum_{k=1}^{\infty} Cap_p(E_k) + \epsilon$$

and letting $\epsilon \to 0^+$, we obtain

$$Cap_p(E) \le \sum_{k=1}^{\infty} Cap_p(E_k).$$

$\square$

**Theorem 4.57** *Let* $A \subset B \subset \mathbb{R}^N$ *and* $1 \le p < N$. *We have the following assertions.*

  (i) $Cap_p(A) = \inf\{Cap_p(U) : A \subset U, U \text{ open set}\}$.

  (ii) $Cap_p(\lambda A) = \lambda^{N-p} Cap_p(A)$, *for every* $\lambda > 0$.

  (iii) $Cap_p(L(A)) = Cap_p(A)$, *for every affine isometry* $L : \mathbb{R}^N \to \mathbb{R}^N$.

  (iv) $Cap_p(B(x,r)) = r^{N-p} Cap_p(B(0,1))$, *for every* $x \in \mathbb{R}^N$, $r > 0$.

  (v) $Cap_p(A) \le CH^{N-p}(A)$ *for some constant* $C \equiv C(N,p)$.

  (vi) $\mathcal{L}^N(A) \le C[Cap_p(A)]^{\frac{N}{N-p}}$ *for some constant* $C \equiv C(N,p)$.

 vii) $Cap_p(A \cap B) + Cap_p(A \cup B) \le Cap_p(A) + Cap_p(B)$.

(viii) *If* $A_1 \subset A_2 \subset \ldots \subset A_k \subset A_{k+1}$, *then*

$$\lim_{k \to +\infty} Cap_p(A_k) = Cap_p(\underset{k \in \mathbb{N}}{\cup} A_k).$$

  (ix) *If* $A_{k+1} \subset A_k \subset \ldots \subset A_2 \subset A_1$ *are compact sets, then*

$$\lim_{k \to +\infty} Cap_p(A_k) = Cap_p(\underset{k \in \mathbb{N}}{\cap} A_k).$$

**Proof**

  (i) It is obvious that

$$Cap_p(A) \le \inf\{Cap_p(U) : A \subset U, U \text{ open set}\}. \tag{4.4}$$

Fix $\epsilon > 0$. There exists $f \in K^p$ such that $A \subset \text{int}\{f \ge 1\} =: U$ and

$$\int_{\mathbb{R}^N} |\nabla f|^p \, dx \le Cap_p(A) + \epsilon.$$

We have

$$Cap_p(U) \le \int_{\mathbb{R}^N} |\nabla f|^p \, dx \le Cap_p(A) + \epsilon.$$

Letting $\epsilon$ go to zero, we deduce that

$$Cap_p(U) \le Cap_p(A)$$

which, together with (4.4), proves assertion (i).

(ii) Fix $\epsilon > 0$. There exists $f \in K^p$ such that $A \subset \text{int}\{f \ge 1\}$ and

$$\int_{\mathbb{R}^N} |\nabla f|^p \, dx \le Cap_p(A) + \epsilon.$$

Setting

$$g(x) := f\left(\frac{x}{\lambda}\right)$$

we have that $g \in K^p$ and $\lambda A \subset \text{int}\{g \ge 1\}$; therefore

$$Cap_p(\lambda A) \le \int_{\mathbb{R}^N} |\nabla g|^p \, dx = \lambda^{N-p} \int_{\mathbb{R}^N} |\nabla f|^p \, dx \le \lambda^{N-p}(Cap_p(A) + \epsilon).$$

Letting $\epsilon$ go to zero, we deduce that

$$Cap_p(\lambda A) \le \lambda^{N-p} Cap_p(A). \tag{4.5}$$

By (4.5) we also have

$$Cap_p(A) = Cap_p\left(\frac{1}{\lambda}(\lambda A)\right) \le \frac{1}{\lambda^{N-p}} Cap_p(\lambda A)$$

and so

$$Cap_p(\lambda A) = \lambda^{N-p} Cap_p(A).$$

(iii) The proof of (iii) uses an argument similar to that of (ii).

(iv) Assertion (iv) is a consequence of assertion (ii). For the proofs of ( v)–(ix), we refer the reader to Evans and Gariepy (1992)

$\square$

The relations between $p$-capacity and Hausdorff measure, as indicated in the previous theorem, can be sharpened, as the following proposition indicates.

**Proposition 4.58** Let $E \subset \mathbb{R}^N$ and $1 < p < N$. If $H^{N-p}(E) < +\infty$, then $Cap_p(E) = 0$. Conversely, if $Cap_p(E) = 0$, then $H^s(E) = 0$ for every $s > N - p$. Moreover, if $p = 1$, then $Cap_1(E) = 0$ if and only if $H^1(E) = 0$.

For the proof, we refer the reader to Evans and Gariepy (1992).

The following result is an immediate consequence of the latter proposition.

**Lemma 4.59** *Let $I \subset (0,1)$ be such that $\mathcal{L}^1(I) > 0$, $N-1 \leq p < N$ and let $E \subset B(0,1) \subset \mathbb{R}^N$. Assume that for each $r \in I$ there exists a unique $x_r \in \partial B(0,r)$ such that $x_r \in E$. Then*

$$Cap_p(E) > 0.$$

**Proof** Let $f : \mathbb{R}^N \to \mathbb{R}$ be defined by $f(x) := |x|_2$. We observe that $f$ is a Lipschitz mapping such that

$$|f(x) - f(y)|_2 \leq |x - y|_2$$

for every $x, y \in \mathbb{R}^N$. Therefore, by Proposition 4.31, we obtain that

$$H^1(f(E)) \leq H^1(E).$$

Together with the fact that $I = f(E)$ and Theorem 4.25, this yields

$$\mathcal{L}^1(I) = H^1(f(E)) \leq H^1(E).$$

If there were $N - 1 < p < N$ such that $Cap_p(E) = 0$, then by Proposition 4.58 and since $1 > N-p$, we would have that $H^1(E) = 0$, which yields a contradiction. Hence,

$$Cap_p(E) > 0.$$

$\square$

Next we show how the notion of $p$-capacity plays an important role in the study of the continuity property for Sobolev functions.

**Definition 4.60** *Let $f : \mathbb{R}^N \to \mathbb{R}$ be a measurable function. $f$ is said to be $p$-quasicontinuous if for every $\epsilon > 0$ there exists an open set $V \subset \mathbb{R}^N$ such that $f|_{\mathbb{R}^N \setminus V}$ is continuous and*

$$Cap_p(V) \leq \epsilon.$$

**Theorem 4.61** *Suppose that $f \in W^{1,p}(\mathbb{R}^N)$ and $1 \leq p < N$.*

(i) *There exists a Borel set $E \subset \mathbb{R}^N$ such that $Cap_p(E) = 0$ and*

$$\lim_{r \to 0^+} \frac{1}{\mathcal{L}^N(B(x,r))} \int_{B(x,r)} f(y)\, dy =: f^*(x)$$

*exists for every $x \in \mathbb{R}^N \setminus E$.*

(ii) *In addition,*

$$\lim_{r \to 0^+} \frac{1}{\mathcal{L}^N(B(x,r))} \int_{B(x,r)} |f(y) - f^*(x)|^{p^*}\, dy = 0$$

*for every $x \in \mathbb{R}^N \setminus E$.*

(iii) $f^*$ *is p-quasicontinuous.*

The proof of this result uses the following lemma.

**Lemma 4.62** *Let* $1 \leq p < N, f \in K^p$, *let* $\epsilon > 0$, *and*

$$E := \left\{ x \in \mathbb{R}^N : \frac{1}{\mathcal{L}^N(B(x,r))} \int_{B(x,r)} f(y)\, dy > \epsilon \text{ for some } r > 0 \right\}.$$

*Then*

$$Cap_p(E) \leq \frac{C}{\epsilon^p} \int_{\mathbb{R}^N} |\nabla f(y)|^p\, dy$$

*for some constant* $C = C(N, p)$.

In turn, this lemma uses the following covering result.

**Theorem 4.63 [Besicovitch's Covering Theorem]** *There exists a constant* $K = K(N)$ *such that if* $\mathcal{F}$ *is any collection of closed balls in* $\mathbb{R}^N$ *and if* $D := \sup\{\text{diam } B : B \in \mathcal{F}\} < +\infty$, *then, if* $A$ *is the set of centres of all balls in* $\mathcal{F}$, *there exist* $\mathcal{G}_1, \ldots, \mathcal{G}_k \subset \mathcal{F}$ *such that* $\mathcal{G}_i$ *is a countable collection of disjoint balls in* $\mathcal{F}$ *and*

$$A \subset \bigcup_{i=1}^{k(N)} \bigcup_{B \in \mathcal{G}_i} B.$$

**Proof of Lemma 4.62** We start by showing that $E$ is an open set. Let $x_0 \in E$ be such that

$$\frac{1}{\mathcal{L}^N(B(x_0,r))} \int_{B(x_0,r)} f(y)\, dy = \epsilon + \alpha$$

for some $\alpha > 0, r > 0$. Choose $0 < \eta$ small enough so that

$$\frac{2}{\mathcal{L}^N(B(x_0,r))} \int_A f(y)\, dy < \alpha$$

whenever $\mathcal{L}^N(A) < \eta$. Finally, let $\delta > 0$ be such that

$$|x - x_0| < \delta \Rightarrow \mathcal{L}^N(B(x,r)\Delta B(x_0,r)) < \eta,$$

where $B\Delta B'$ denotes the symmetric difference between $B$ and $B', B\Delta B' = B\backslash B' \cup B'\backslash B$.

Then, if $|x - x_0| < \delta$ we have

$$\frac{1}{\mathcal{L}^N(B(x,r))} \int_{B(x,r)} f(y)\,dy = \frac{1}{\mathcal{L}^N(B(x_0,r))} \int_{B(x_0,r)} f(y)\,dy$$

$$+ \frac{1}{\mathcal{L}^N(B(x_0,r))} \left( \int_{B(x,r)} f(y)\,dy - \int_{B(x_0,r)} f(y)\,dy \right)$$

$$\geq \alpha + \epsilon - \frac{2}{\mathcal{L}^N(B(x_0,r))} \int_{B(x,r)\Delta B(x_0,r)} f(y)\,dy$$

$$> \epsilon$$

and this proves that $E$ is open. Note that if $x_0 \in E$ and if $r$ satisfies

$$\frac{1}{\mathcal{L}^N(B(x_0,r)} \int_{B(x_0,r)} f(y)\,dy > \epsilon$$

then

$$\alpha(N)r^N \epsilon \leq \int_{B(x_0,r)} f(y)\,dy$$

$$\leq [\alpha(N)r^N]^{1-1/p*} \left( \int_{B(x_0,r)} f(y)^{p^*}\,dy \right)^{1/p^*}$$

$$\leq [\alpha(N)r^N]^{1-1/p^*} \left( \int_{\mathbb{R}^N} f(y)^{p^*}\,dy \right)^{1/p^*}$$

and so there exists a constant $C_0$ independent of $x_0$ such that

$$r \leq C_0. \tag{4.6}$$

By Besicovitch's Covering Theorem there exists $K = K(N)$ and a collection of disjoint closed balls $\mathcal{G}_1,\ldots,\mathcal{G}_k$ such that $E \subset \cup_{i=1}^k \cup_{B\in\mathcal{G}_i} B$ and

$$\frac{1}{\mathcal{L}^N(B)} \int_B f(y)\,dy > \epsilon \tag{4.7}$$

for every $B \in \cup_{i=1}^k \mathcal{G}_i$. Let $\mathcal{G}_i = \{B_j^{(i)}\}_{j=1}^\infty$. By Hölder's Inequality and Theorem 4.42 we have

$$\left[ \frac{1}{\mathcal{L}^N(B_j^{(i)})} \int_{B_j^{(i)}} f(y)\,dy - f \right]^+ \in W^{1,p}(B_j^{(i)})$$

and, due to Poincaré's Inequality,

$$\left\| \left( \frac{1}{\mathcal{L}^N(B_j^{(i)})} \int_{B_j^{(i)}} f(y)\,dy - f \right)^+ \right\|_{W^{1,p}(B_j^{(i)})} \leq C_1 \, \| \nabla f \|_{L^p(B_j^{(i)})} \cdot$$

Moreover, by the Extension Theorem 4.47 and Remark 4.48, there exist a constant $C_2$ and $h_j^{(i)} \in W^{1,p}(\mathbb{R}^N)$ such that

$$h_j^{(i)} \geq 0, \quad h_j^{(i)}(x) = \left[ \frac{1}{\mathcal{L}^N(B_j^{(i)})} \int_{B_j^{(i)}} f(y)\,dy - f(x) \right]^+ \quad \text{a.e. in } B_j^{(i)}$$

and

$$\| h_j^{(i)} \|_{W^{1,p}(\mathbb{R}^N)} \leq C_2 \left\{ \int_{B_j^{(i)}} |\nabla f(x)|^p \, dx \right\}^{\frac{1}{p}}, \tag{4.8}$$

where, due to (4.6), $C_2$ is independent of $B_j^{(i)}$. Then, by (4.7),

$$f + h_j^{(i)} \geq \frac{1}{\mathcal{L}^N(B_j^{(i)})} \int_{B_j^{(i)}} f(y)\,dy \geq \epsilon \quad \text{a.e. in } B_j^{(i)}$$

and, setting

$$h := \sup\{h_j^{(i)} : \ i = 1, \ldots, k, j \in \mathbb{N}\},$$

we have $h \geq 0$. We claim that $h \in K^p$. Indeed, $h \in L^p(\mathbb{R}^N)$ because, by (4.8),

$$\int_{\mathbb{R}^N} \sup_{i,j}[h_j^{(i)}(x)]^p \, dx \leq \sum_{i=1}^{k} \sum_{j=1}^{\infty} \int_{\mathbb{R}^N} [h_j^{(i)}(x)]^p \, dx$$

$$\leq \sum_{i=1}^{k} C_2^p \sum_{j=1}^{\infty} \int_{B_j^{(i)}} |\nabla f(x)|^p \, dx$$

$$\leq K C_2^p \int_{\mathbb{R}^N} |\nabla f(x)|^p \, dx.$$

Also, by (4.8),

$$\int_{\mathbb{R}^N} \sup_{i,j} |\nabla h_j^{(i)}(x)|^p \, dx \leq \sum_{i=1}^{k} \sum_{j=1}^{\infty} \int_{\mathbb{R}^N} |\nabla h_j^{(i)}(x)|^p \, dx$$

$$\leq \sum_{i=1}^{k} C_2^p \sum_{j=1}^{\infty} \int_{B_j^{(i)}} |\nabla f(x)|^p \, dx$$

$$\leq K C_2^p \int_{\mathbb{R}^N} |\nabla f(x)|^p \, dx. \tag{4.9}$$

By Proposition 4.43, we conclude that $h \in W^{1,p}(\mathbb{R}^N)$ and

$$|\nabla h(x)| \leq \sup_{i,j} |\nabla h_j^{(i)}(x)|^p \, dx \quad \text{a.e. } x \in \mathbb{R}^N. \tag{4.10}$$

Due to the Sobolev–Niremberg–Gagliardo Inequality (see Theorem 4.44), $h \in K^p$, and as $f + h \geq \epsilon$ a.e. in $E$ and as $E$ is an open set we deduce that

$$Cap_p(E) \leq \int_{\mathbb{R}^N} \left| \nabla \left( \frac{f+h}{\epsilon} \right)(x) \right|^p dx$$

$$\leq \frac{C}{\epsilon^p} \left\{ \int_{\mathbb{R}^N} |\nabla f(x)|^p \, dx + \int_{\mathbb{R}^N} |\nabla h(x)|^p \, dx \right\},$$

which, together with (4.9) and (4.12), yields

$$Cap_p(E) \leq \frac{C}{\epsilon^p} \int_{\mathbb{R}^N} |\nabla f(x)|^p \, dx.$$

$\square$

**Remark 4.64**

(i) Theorem 4.61 asserts that if $1 \leq p < N$, then, up to a set $E$ of $p$-capacity zero, a function $f \in W^{1,p}(\mathbb{R}^N)$ can be represented by a $p$-quasicontinuous function. In particular, by Proposition 4.58 we conclude that

$$H^s(E) = 0 \text{ for every } s > N - p$$

and so

$$H_{dim}(E) \leq N - p. \tag{4.11}$$

On the other hand, (4.11) does not necessarily imply that $H^{N-p}(E) < +\infty$, which, in turn, yields

$$Cap_p(E) = 0. \tag{4.12}$$

by Proposition 4.58. Therefore, (4.12) may be a stronger statement than (4.11).

(ii) Recently, Malý sharpened Theorem 4.61 by showing that the quasicontinuous representative $f^*$ of $f \in W^{1,N}(\Omega; \mathbb{R}^N)$ is approximately Hölder continuous, except on a set of Hausdorff dimension zero, i.e. if $S$ is the set of all points of $\Omega$ at which $f^*$ is approximately Hölder continuous, then

$$H_{dim}(\Omega \backslash S) = 0.$$

Actually, in view of (i), Malý's precise statement asserts that for every $\epsilon > 0, p < N$, there exists an open set $G \subset \mathbb{R}^N$ such that $\text{Cap}_p(G) < \epsilon$ and $f|_{\Omega \backslash G}$ is locally Hölder continuous.

Here $f^*$ is said to be *approximately Hölder continuous* at $x_0$ if for every $\alpha \in (0,1]$ there exists a set $M$ such that the *Lebesgue density* of $M$ at $x_0$ is 1, i.e.

$$\lim_{\epsilon \to 0^+} \frac{\mathcal{L}^N(M \cap B(x_0, \epsilon))}{\mathcal{L}^N(B(x_0, \epsilon))} = 1$$

and

$$\limsup_{y \to x_0, y \in M} \frac{|f^*(y) - f^*(x)|}{|y - x_0|^\alpha} < +\infty.$$

**Proof of Theorem 4.61** Let $f \in W^{1,p}(\mathbb{R}^n), 1 \le p < N$.

(i) We want to show that there exists $E \subset \mathbb{R}^N$, $Cap_p(E) = 0$ such that

$$\lim_{r \to 0^+} \frac{1}{\mathcal{L}^N(B(x,r))} \int_{B(x,r)} f(y)\, dy =: f^*(x)$$

for every $x \in \mathbb{R}^N \setminus E$. We define

$$A := \left\{ x \in \mathbb{R}^N : \limsup_{r \to 0^+} \frac{1}{r^{N-p}} \int_{B(x,r)} |\nabla f(y)|^p\, dy > 0 \right\}.$$

By Proposition 4.37, $H^{N-p}(A) = 0$ and so, by Theorem 4.57 (v),

$$Cap_p(A) = 0.$$

By Poincaré's Inequality we have

$$\lim_{\epsilon \to 0^+} \frac{1}{\mathcal{L}^N(B(x,r))} \int_{B(x,r)} |f(y) - \frac{1}{\mathcal{L}^N(B(x,r))} \int_{B(x,r)} f(z) dz|^{p^*}\, dy$$

$$\le C \lim_{r \to 0^+} \frac{1}{r^N} \left\{ \int_{B(x,r)} |\nabla f(y)|^p\, dy \right\}^{p^*/p}$$

$$= C \lim_{r \to 0^+} \left\{ \frac{1}{r^{N-p}} \int_{B(x,r)} |\nabla f(y)|^p\, dy \right\}^{p^*/p}$$

$$= 0 \tag{4.13}$$

for every $x \notin A$. Due to the density of smooth functions in $W^{1,p}$, for each $i \in \mathbb{N}$ we choose $f_i \in W^{1,p}(\mathbb{R}^N) \cap C^\infty(\mathbb{R}^N)$ such that

$$\int_{\mathbb{R}^N} |\nabla f(x) - \nabla f_i(x)|^p\, dx \le \frac{1}{2^{(p+1)i}}$$

and we set

$$B_i := \left\{ x \in \mathbb{R}^N : \frac{1}{\mathcal{L}^N(B(x,r))} \int_{B(x,r)} |f(y) - f_i(y)|\, dy > \frac{1}{2^i}, \right.$$
$$\left. \text{for some } r > 0 \right\}.$$

By Lemma 4.62 and also because, due to Theorem 4.44, if $f \in W^{1,p}(\mathbb{R}^N)$, then $|f| \in K^p$, we have

$$\frac{Cap_p(B_i)}{2^{pi}} \leq C \int_{\mathbb{R}^N} |\nabla f(x) - \nabla f_i(x)|^p \, dx$$

$$\leq \frac{C}{2^{(p+1)i}},$$

i.e.

$$Cap_p(B_i) \leq \frac{C}{2^i}. \tag{4.14}$$

Now

$$\left| \frac{1}{\mathcal{L}^N(B(x,r))} \int_{B(x,r)} f(y) \, dy - f_i(x) \right|$$

$$\leq \frac{1}{\mathcal{L}^N(B(x,r)} \int_{B(x,r)} \left| f(y) - \frac{1}{\mathcal{L}^N(B(x,r))} \int_{B(x,r)} f(z) \, dz \right| dy$$

$$+ \frac{1}{\mathcal{L}^N(B(x,r)} \int_{B(x,r)} |f(z) - f_i(z)| \, dz$$

$$+ \frac{1}{\mathcal{L}^N(B(x,r))} \int_{B(x,r)} |f_i(z) - f_i(x)| \, dz.$$

As $f_i$ is smooth, the last integral in the above inequality converges to zero as $r \to 0^+$, and so, due to (4.13), we conclude that

$$\limsup_{r \to 0^+} \left| \frac{1}{\mathcal{L}^N(B(x,r))} \int_{B(x,r)} f(y) \, dy - f_i(x) \right| \leq \frac{1}{2^i} \tag{4.15}$$

for every $x \notin (A \cup B_i)$. Set

$$E_k := A \cup \left( \overset{\infty}{\underset{j=k}{\cup}} B_j \right).$$

Then, by Theorem 4.56 and (4.14), we have

$$Cap_p(E_k) \leq Cap_p(A) + \sum_{j=k}^{+\infty} Cap_p(B_j)$$

$$\leq C \sum_{j=k}^{+\infty} \frac{1}{2^j}$$

and if $x \in \mathbb{R}^N \backslash E_k, i, j \geq k$, then, by (4.15),

$$|f_i(x) - f_j(x)| \leq \limsup_{r \to 0+} \left| \frac{1}{\mathcal{L}^N(B(x,r))} \int_{B(x,r)} f(y)\, dy - f_i(x) \right|$$

$$+ \limsup_{r \to 0+} \left| \frac{1}{\mathcal{L}^N(B(x,r))} \int_{B(x,r)} f(y)\, dy - f_j(x) \right|$$

$$\leq \frac{1}{2^i} + \frac{1}{2^j}.$$

We conclude that $f_i \to g$ in $L^\infty(\mathbb{R}^N \setminus E_k)$ and $g$ is a continuous function. Also,

$$\limsup_{r \to 0+} \left| g(x) - \frac{1}{\mathcal{L}^N(B(x,r))} \int_{B(x,r)} f(y)\, dy \right|$$

$$\leq |g(x) - f_i(x)| + \limsup_{r \to 0+} \left| f_i(x) - \frac{1}{\mathcal{L}^N(B(x,r))} \int_{B(x,r)} f(y)\, dy \right|$$

$$\leq |g(x) - f_i(x)| + \frac{1}{2^i},$$

where we have used (4.15). Thus,

$$g(x) = \lim_{r \to 0+} \frac{1}{\mathcal{L}^N(B(x,r))} \int_{B(x,r)} f(y)\, dy$$
$$=: f^*(x)$$

for every $x \in \mathbb{R}^N \setminus E_k$. Setting $E := \cap_{k \in \mathbb{N}} E_k$, then, by Theorem 4.57,

$$Cap_p(E) \leq \lim_{k \to +\infty} Cap_p(E_k)$$
$$= 0$$

and $f^*(x)$ exists for every $x \in \mathbb{R}^N \setminus E$.

(ii) We claim that

$$\lim_{r \to 0+} \frac{1}{\mathcal{L}^N(B(x,r))} \int_{B(x,r)} |f(y) - f^*(x)|^{p^*}\, dy = 0.$$

Indeed, for every $x \notin E$, by definition of $f^*$ we have

$$\lim_{r \to 0+} \left\{ \frac{1}{\mathcal{L}^N(B(x,r))} \int_{B(x,r)} |f(y) - f^*(x)|^{p^*}\, dy \right\}^{1/p^*}$$

$$\leq \lim_{r \to 0+} \left\{ \frac{1}{\mathcal{L}^N(B(x,r))} \int_{B(x,r)} \left| f(y) - \frac{1}{\mathcal{L}^N(B(x,r))} \int_{B(x,r)} f(z)\, dz \right|^{p^*}\, dy \right.$$

$$+ \frac{1}{\mathcal{L}^N(B(x,r))} \int_{B(x,r)} \left| \frac{1}{\mathcal{L}^N(B(x,r))} \int_{B(x,r)} f(z)\, dz - f^*(x) \right|^{p^*} dy \Bigg\}^{1/p^*}$$

$$= \lim_{r \to 0+} \left| \frac{1}{\mathcal{L}^N(B(x,r))} \int_{B(x,r)} f(z)\, dz - f^*(x) \right|$$

$$= 0.$$

(iii) It remains to prove that $f^*$ is $p$-quasicontinuous. Fix $\epsilon > 0$ and choose $k$ large enough so that $Cap_p(E_k) < \frac{\epsilon}{2}$. By Theorem 4.57 (i) there exists an open set $U \supset E_k$ such that $Cap_p(U) < \epsilon$. Then

$$f_i \to f^* \text{ in } L^\infty(\mathbb{R}^N \backslash U)$$

and so $f^* \mid_{(\mathbb{R}^N \backslash U)}$ is continuous.

□

There are other notions of capacity, such as *Bessel capacity, linear capacity,* etc., and the interested reader may find a detailed description in Hayman and Kennedy (1976), Havin and Maz'ya (1972), Stein (1970), and Ziemer (1989). With the help of these concepts one can prove the following.

**Theorem 4.65** *Let $f \in L^1(\mathbb{R}^N)$ be a function with compact support, spt $f \subset K \subset\subset \mathbb{R}^N$. Let*

$$w(x) := \int_k \frac{f(y)}{|x - y|^\alpha}\, dy,$$

*where $\alpha \in [0,1)$, and let $K_1 \subset \mathbb{R}^N$ be a compact set such that $w(x) = +\infty$ on $K_1$. Then $Cap_{N-\alpha-\epsilon}(K_1) = 0$ for all $\alpha > \epsilon > 0$.*

The proof of this result may be found in Hayman and Kennedy (1976), where it is stated more generally for measures of bounded variation in place of $\mathcal{L}^N$.

# 5

# PROPERTIES OF THE DEGREE FOR SOBOLEV FUNCTIONS

In this chapter we present a generalization of Sard's Lemma for Sobolev functions and the following change of variables formulae. Let $D \subset \mathbb{R}^N$ be open and bounded and assume that either $p > N$ or $N - 1 < p \le N$ and $J_\phi(x) > 0$ $\mathcal{L}^N$ a.e. $x \in D$. Then

$$\int_D v \circ \phi(x) |J_\phi(x)| \, dx = \int_{\mathbb{R}^N} v(y) N(\phi, D, y) \, dy$$

for every $v \in L^\infty(\mathbb{R}^N)$ and for every $\phi \in W^{1,p}(D)^N$, where the *multiplicity function of $\phi$ at $y \in \mathbb{R}^N$ with respect to $D$* is defined by

$$N(\phi, D, y) := \sharp \{x \in D : \phi(x) = y\}.$$

A second change of variables formula is

$$\int_D v \circ \phi(x) J_\phi(x) \, dx = \int_{\mathbb{R}^N} v(y) d(\phi, D, y) \, dy$$

for every $v \in L^\infty(\mathbb{R}^N)$ and for every $\phi \in W^{1,p}(D)^N$, where $d(\phi, D, y)$ stands for the topological degree of $\phi$ with respect to $D$ at the point $y$.

As an immediate consequence of the latter change of variables formula, if a sequence $\{\phi_n\} \subset W^{1,p}(D)^N$ converges to $\phi$ uniformly, then the sequence $\{d(\phi_n, D, y)\} \subset \mathbb{Z}$ converges to $d(\phi, D, y)$ and so, under some additional assumptions, $\{\int_D v \circ \phi_n(x) J_{\phi_n}(x) \, dx\}$ converges to $\int_D v \circ \phi(x) J_\phi(x) \, dx$.

Given an open set $D \subset \mathbb{R}^N$, we will use the following notation: if $1 \le p \le +\infty$, $f \in L^p(D)$, then

$$||f||_p^p(D) := \int_D |f|^p \, dx, \quad \text{if } p < \infty$$

and

$$||f||_\infty(D) := \inf \{M \ge 0 : |f(x)| \le M \ \mathcal{L}^N \text{ a.e. } x \in D\}, \quad \text{if } p = \infty.$$

If $\phi \in L^p(D)^M$, then

$$||\phi||_p^p(D) := \sum_{i=1}^N \int_D |\phi_i|^p$$

and if $\phi \in W^{1,p}(D)^M$, then

$$||\phi||_{1,p}^p(D) := ||\phi||_p^p(D) + \sum_{i,j=1}^N \left|\left|\frac{\partial \phi_i}{\partial x_j}\right|\right|_p^p(D).$$

## 5.1   Results of weakly differential mappings

In this section we present some results relating the notion of topological degree to weak differentiability conditions. As usual, we denote by $B$ the unit ball centred at 0 of $\mathbb{R}^N$ with respect to the norm $|\cdot|_2$.

**Lemma 5.1** *Let $F \in C(B)^N$ and let $L$ be a linear mapping of $\mathbb{R}^N$ into $\mathbb{R}^N$. Then*

$$\mathcal{L}^N(F(P)) \leq \mathcal{L}^N(L(B)) + O(\epsilon),$$

*where $P := \{x \in \bar{B} : |F(x) - L(x)| \leq \epsilon\}$.*

**Proof** Fix $\epsilon > 0$. If $\det(L) = 0$, then $L$ maps $\bar{B}$ into a hyperplane $P$ of dimension $N-1$. Let

$$l := \max\{|L(x)| : x \in \bar{B}\}.$$

Then $F(P)$ lies inside a parallelepiped which has $(N-1)$ sides of length less than $2l + \epsilon$ and the $N$th side of length less than $2\epsilon$. Hence,

$$\mathcal{L}^N(F(P)) \leq (2l + \epsilon)^{N-1}2\epsilon = \mathcal{L}^N(L(B)) + O(\epsilon) \tag{5.1}$$

because $\mathcal{L}^N(L(B)) = 0$. If $\det(L) \neq 0$, then $L$ is a homeomorphism of $\mathbb{R}^N$ onto $\mathbb{R}^N$ and

$$L^{-1}(F(P)) \subset \{x \in \mathbb{R}^N : \rho(x, \bar{B}) \leq \epsilon||L^{-1}||\},$$

where we recall that $\rho(x, \bar{B}) = \inf\{|x - t| : t \in \bar{B}\}$ and $||L^{-1}||$ stands for the norm of $L^{-1}$ in $C(\partial B)$. Thus,

$$\mathcal{L}^N(L^{-1}(F(P))) \leq w_N(1 + \epsilon||L^{-1}||)^N,$$

where $w_N := \mathcal{L}^N(B)$. Since $L \in C^\infty(\mathbb{R}^N)^N$, by the usual change of variables formula we obtain that

$$\mathcal{L}^N((F(P))) \leq w_N(1 + \epsilon||L^{-1}||)^N\det(L)$$

and so

$$\mathcal{L}^N((F(P))) \leq \mathcal{L}^N(L(B)) + O(\epsilon).$$

$\square$

**Definition 5.2** *Let $D \subset \mathbb{R}^N$ be an open set and let $\phi : D \to \mathbb{R}^N$ be a continuous function. For $x \in D$ we set*

$$L_h(y) := \frac{\phi(x + hy) - \phi(x)}{h},$$

*where $X \in B$. Let $L : \mathbb{R}^N \to \mathbb{R}^N$ be a linear mapping.*

(i) $L$ is called the approximate differential of $\phi$ at $x$ if $L_h$ converges to $L$ in measure on the ball $B$. We write

$$L = (app)d\phi_x.$$

(ii) $L$ is called the weak differential of $\phi$ at $x$ if $L = (app)d\phi_x$ and if there exists a sequence $\{h_m\}$ converging to $0$ when $m$ tends to infinity such that $\{L_{h_m}\}$ converges to $L$ uniformly on the sphere $S^{N-1} := \partial B$.

(iii) If $\phi$ has a weak differential at $\mathcal{L}^N$ almost every point of $D$, we say that $\phi$ is weakly differentiable on $D$.

**Definition 5.3** Let $D \subset \mathbb{R}^N$ be an open set and let $\phi : D \to \mathbb{R}^N$.

(i) We say that $\phi$ has the $N$-property if

$$\mathcal{L}^N(\phi(E)) = 0$$

for every $E \subset D$ such that $\mathcal{L}^N(E) = 0$.

(ii) We say that $\phi$ has the $N^{-1}$-property if

$$\mathcal{L}^N(\phi^{-1}(F)) = 0$$

for every $F \subset \mathbb{R}^N$ measurable set such that $\mathcal{L}^N(F) = 0$.

Functions verifying the $(N)$ property map measurable sets into measurable sets. Precisely,

**Lemma 5.4** Let $D \subset \mathbb{R}^N$ be an open set and let $\phi \in C(D)^N$ be a function satisfying the $N$-property. If $E \subset D$ is measurable, then $\phi(E)$ is also measurable.

**Proof**  Since $E$ is measurable there exists a sequence $\{K_n : n \in \mathbb{N}\}$ of compact sets such that

$$K_n \subset K_{n+1}$$

and

$$\mathcal{L}^N(E \setminus K_n) \le \frac{1}{n}.$$

We have

$$E = \underset{n\in\mathbb{N}}{\cup}\, K_n \cup N,$$

where $N \subset D$ is a set of measure zero; hence

$$\phi(E) = \underset{n\in\mathbb{N}}{\cup}\, \phi(K_n) \cup \phi(N).$$

Since $\phi$ is a continuous function, the sets $\phi(K_n)$ are compact sets, hence measurable, and since $\mathcal{L}^N(N) = 0$, and $\phi$ satisfies the $N$-property, we have

$$\mathcal{L}^N(\phi(N)) = 0.$$

Thus $\phi(N)$ is measurable and $\phi(E)$ is measurable because it is a countable union of measurable sets.    □

**Theorem 5.5** *Let $D \subset \mathbb{R}^N$ be an open set and let $\phi \in C(D)^N$ be a mapping which has the N-property. Then $y \to N(\phi, E, y)$ is measurable, for every $E \subset D$ that is measurable.*

**Proof** Since

$$\lim_{k \to +\infty} N(\phi, E_k, y) = N(\phi, E, y)$$

for every $y \in \mathbb{R}^N$ and for every sequence of measurable sets $\{E_k : k \in \mathbb{N}\}$ such that

$$E = \underset{k \in \mathbb{N}}{\cup} E_k \quad \text{and} \quad E_1 \subset E_2 \subset \ldots \subset E_k \subset E_{k+1},$$

we may assume without loss of generality that $E$ is bounded. We first prove that

$$\lim_{m \to +\infty} N(\phi, m, y) = N(\phi, E, y),$$

where

$$N(\phi, m, y) = \chi_{\phi(E_1^m)} + \cdots + \chi_{\phi(E_{k(m)}^m)},$$

and $\{E_1^m, \ldots, E_{k(m)}^m\}$ is any measurable, pairwise nonintersecting partition of $E$ such that $\mathrm{diam} E_i^m) \leq \frac{1}{m}$, $i = 1, \ldots, k(m)$. It is easy to see that we have

$$N(\phi, m, y) \leq N(\phi, E, y). \tag{5.2}$$

If $\phi^{-1}\{y\} \cap E = \{a_1, \ldots, a_l\}$, we fix

$$m_0 > \frac{1}{\underset{i \neq j}{\min} |a_i - a_j|_2}.$$

We observe that if $m \geq m_0$ and if $\mathrm{diam}(E_i^m) \leq \frac{1}{m}$, $i = 1, \ldots, k(m)$, none of the $E_i^m$ contains two differents $a_j$, $a_r$ and so we assume without loss of generality that

$$a_1 \in E_1^m, \ldots, a_l \in E_l^m.$$

It is obvious that

$$N(\phi, m, y) = l \geq N(\phi, E, y),$$

which, together with (5.2), yields

$$\lim_{m \to \infty} N(\phi, m, y) = N(\phi, E, y).$$

If $\sharp(\phi^{-1}\{y\} \cap E) = +\infty$, then using the argument above, we obtain that for every $l \in \mathbb{N}$ there exists $m \in \mathbb{N}$ such that

$$N(\phi, m, y) \geq l$$

and so, by (5.2),

$$\lim_{m \to \infty} N(\phi, m, y) = N(\phi, E, y).$$

Since $E$ is a measurable, bounded set, for every $m \in \mathbb{N}$ there exists $k(m) \in \mathbb{N}$ and a pairwise nonintersecting partition of $E$, $\left\{ E_1^m, \ldots, E_{k(m)}^m \right\}$, such that

$$\operatorname{diam}(E_i^m) \le \frac{1}{m}, \quad i = 1, \ldots, k(m).$$

By Lemma 5.4, $\phi(E_i^m)$ is measurable for every $i = 1, \ldots, k(m)$ and $N(\phi, m, \cdot)$ is measurable. Therefore, $N(\phi, E, \cdot)$ is measurable since it can be written as a limit of measurable functions. $\qquad \square$

We present a generalization of Sard's Lemma for Sobolev functions and the proof we give is due to Gold'sthein and Reshetnyak (1990).

**Theorem 5.6** *Let $D \subset \mathbb{R}^N$ be an open set and let $\phi \in C(D)^N$ be a mapping which has the $N$-property. Assume that $\phi$ has an approximate differential almost everywhere and that the Jacobian $J_\phi$ is locally summable in $D$. Then for every measurable set $E \subset D$ we have*

$$\mathcal{L}^N(\phi(E)) \le \int_E |J_\phi(x)| \, dx. \tag{5.3}$$

**Remark 5.7**

(i) The proof of Theorem 5.6 was given by Schwartz (1969) in the case where $\phi \in C^1(\bar{D})^N$.

(ii) If $\phi \in W^{1,N}(D)^N$, $J_\phi(x) > 0$ $\mathcal{L}^N$ a.e. $x \in D$, we will prove that $\phi$ satisfies the assumptions of Theorem 5.6 (see Theorems 5.17, 5.21, and 5.32).

**Proof of Theorem 5.6.** Clearly, it suffices to consider the case where $E$ is bounded. Let $M_1$ be the set of points in $D$ where the approximate differential does not exist and let $M_2$ be the set of points which are not Lebesgue points of $|J_\phi|$. Since $\mathcal{L}^N(M_1) = \mathcal{L}^N(M_2) = 0$ and $\phi$ has the $N$-property, we have

$$\mathcal{L}^N(\phi(E)) = \mathcal{L}^N(\phi(E \setminus (M_1 \cup M_2))).$$

Hence, in the sequel we may assume without loss of generality that $E \cap (M_1 \cup M_2) = \emptyset$ and

$$\int_E |J_\phi(x)| \, dx < +\infty.$$

*Case 1.* Suppose that $\bar{E} \subset D$.

Fix an arbitrary $\epsilon > 0$ and let $G_1$ be an open set such that

$$E \subset\subset G_1 \subset\subset D$$

and

$$\mathcal{L}^N(G_1) \leq \mathcal{L}^N(E) + \frac{\epsilon}{2}.$$

Define

$$\eta(A) := \int_A |J_\phi(x)|\, dx,$$

for every measurable set $A \subset G_1$. Since $J_\phi \in L^1(G_1)$, there is $\delta > 0$ such that

$$\mathcal{L}^N(A) \leq \delta \Rightarrow \eta(A) < \frac{\epsilon}{2}, \tag{5.4}$$

for every measurable $A \subset G_1$. Let $G$ be an open set such that

$$E \subset\subset G \subset\subset G_1$$

and

$$\mathcal{L}^N(G \setminus E) \leq \delta. \tag{5.5}$$

Let $x_0 \in E$, $L = (app)d\phi_{x_0}$ and

$$L_h(y) := \frac{\phi(x_0 + hy) - \phi(x_0)}{h}, \quad y \in B.$$

Set

$$\tau := \frac{1}{2} \min \left\{ \frac{\delta}{\mathcal{L}^N(E) + \delta}, \frac{\epsilon}{\mathcal{L}^N(E) + \delta} \right\},$$

$$Q(h) := \{y \in \bar{B} : |L_h(y) - L(y)| > h_0\},$$

$$P(h) := \bar{B} \setminus Q(h).$$

Since $\{L_h\}$ converges to $L$ in measure when $h$ tends to zero, there exists $0 < h_1 \leq h_0$ such that

$$\mathcal{L}^N(Q(h)) < \tau \mathcal{L}^N(B) \tag{5.6}$$

for every $0 < h \leq h_1$ and by Lemma 5.1 there exists $0 < h_2 < h_1$ such that

$$\mathcal{L}^N(L_h(P(h))) \leq \mathcal{L}^N(L(B)) + \mathcal{L}^N(B)\tau \tag{5.7}$$

for every $0 < h \leq h_2$. Using the fact that $x_0$ is a Lebesgue point for $J_\phi$, we deduce that

$$\mathcal{L}^N(B(x_0, h))|J_\phi(x_0)| < \int_{B(x_0, h)} J_\phi(y)\, dy + \tau \mathcal{L}^N(B(x_0, h)) \tag{5.8}$$

for some $0 < h_3 \leq h_2$ and for every $0 < h \leq h_3$. Consider the mappings $F : \mathbb{R}^N \to \mathbb{R}^N$ and $S : \mathbb{R}^N \to \mathbb{R}^N$ defined by

$$F(y) := x_0 + hy, \quad S(y) := \phi(x_0) + hy.$$

Set

$$Q(x_0, h) := F(Q(h)), \quad P(x_0, h) := F(P(h)).$$

It is obvious that

$$Q(x_0, h) \subset \bar{B}(x_0, h)$$

and

$$P(x_0, h) = \bar{B}(x_0, h) \setminus Q(x_0, h),$$

which, together with (5.6), implies that

$$\mathcal{L}^N(Q(x_0, h)) < \tau h^N \mathcal{L}^N(B(0, 1)) = \tau \mathcal{L}^N(B(x_0, h)) \tag{5.9}$$

for every $0 < h \leq h_3$. Using the above definition of $L$ and $L_h$, we have $\phi \circ F = S \circ L_h$ and so

$$\phi((P(x_0, h)) = S \circ L_h(P(h)); \tag{5.10}$$

hence

$$\mathcal{L}^N(\phi(P(x_0, h))) = h^N \mathcal{L}^N(L_h(P(h))), \tag{5.11}$$

which, together with (5.7), yields

$$\mathcal{L}^N(\phi(P(x_0, h))) \leq \mathcal{L}^N(L(B(x_0, h))) + \tau \mathcal{L}^N(B(x_0, h)). \tag{5.12}$$

Recalling that $\det(L) = J_\phi(x_0)$, we have

$$\mathcal{L}^N(L(B(x_0, h))) = |J_\phi(x_0)| \mathcal{L}^N(B(x_0, h))$$

and by (5.8) and (5.12) we deduce that

$$\mathcal{L}^N(\phi(P(x_0, h))) \leq \int_{B(x_0, h)} |J_\phi(x)| \, dx + 2\tau \mathcal{L}^N(B(x_0, h)) \tag{5.13}$$

for every $0 < h \leq h_3$. Let $h_4(x_0) := \mathrm{dist}(x_0, \partial G)$. Since $\bar{E} \subset G$ we obtain that $h_4(x_0) > 0$ and we set

$$h(x_0) := \min\{h_3, h_4(x_0)\}.$$

Using a corollary of Vitali's Covering Theorem (see Corollary 4.35), we obtain the existence of a countable sequence $\{x_n\}_{n \in \mathbb{N}} \subset E$ and a sequence $0 < h_n < h(x_n)$ such that the balls $B(x_n, h_n)$ are mutually disjoint and

$$\mathcal{L}^N(E \setminus \bigcup_{n \in \mathbb{N}} B(x_n, h_n)) = 0. \tag{5.14}$$

Set

$$T := \bigcup_{n \in \mathbb{N}} B(x_n, h_n), \quad P := \bigcup_{n \in \mathbb{N}} P(x_n, h_n) \text{ and } Q := \bigcup_{n \in \mathbb{N}} Q(x_n, h_n).$$

By (5.13) we obtain that

$$\mathcal{L}^N(\phi(P \cap E)) \le \sum_{n=1}^{\infty} \mathcal{L}^N(\phi(P(x_n, h_n)))$$

$$\le \int_G |J_\phi(x)|\, dx + 2\tau \mathcal{L}^N(G)$$

$$\le \int_G |J_\phi(x)|\, dx + \min\{\epsilon, \delta\}.$$

By (5.4) and (5.5) we deduce that

$$\mathcal{L}^N(\phi(P \cap E)) \le \int_E |J_\phi(x)|\, dx + 2\epsilon. \tag{5.15}$$

Going back to (5.9), we obtain that

$$\mathcal{L}^N(Q) \le \sum_{n=1}^{\infty} \mathcal{L}^N(Q(x_n, h_n)) \le \tau \mathcal{L}^N(G) \le \min\{\epsilon, \delta\}$$

which, together with (5.4) and (5.15), yields

$$\mathcal{L}^N(\phi(P \cap E)) \le \int_{E \cap P} |J_\phi(x)|\, dx + 3\epsilon.$$

Set $E_0 = E$ and $E_1 = Q$. We proved that if $E_k \subset\subset D$, then there exists a measurable set $E_{k+1} \subset\subset E_k$ such that

$$\mathcal{L}^N(E_{k+1}) \le \frac{\epsilon}{2^k}$$

and

$$\mathcal{L}^N(\phi(E_k \setminus E_{k+1})) \le \int_{E_k \setminus E_{k+1}} |J_\phi(x)|\, dx + \frac{3\epsilon}{2^k}. \tag{5.16}$$

Let $N := \cap_{k \in \mathbb{N} \cup \{0\}} E_k$. Then $N$ is a set of measure zero and we observe that

$$E = N \cup \bigcup_{k \in \mathbb{N} \cup \{0\}} (E_k \setminus E_{k+1}).$$

Using (5.16) and the fact that $\phi$ has the N-property, we have

$$\mathcal{L}^N(\phi(E)) = \mathcal{L}^N(\phi(\bigcup_{k \in \mathbb{N} \cup \{0\}} (E_k \setminus E_{k+1})))$$

$$\le \sum_{k=0}^{\infty} \mathcal{L}^N(\phi(E_k \setminus E_{k+1}))$$

$$\le \int_{E \setminus N} |J_\phi(x)|\, dx + 12\epsilon$$

$$= \int_E |J_\phi(x)| \, dx + 12\epsilon.$$

Letting $\epsilon$ tend to zero, we conclude that

$$\mathcal{L}^N(\phi(E)) \le \int_E |J_\phi(x)| \, dx. \tag{5.17}$$

*Case 2.* We assume that $E \subset D$.

Since $E$ is measurable there exists a sequence of compact sets $\{K_n : n \in \mathbb{N}\}$ such that

$$K_n \subset K_{n+1} \subset E \text{ and } \mathcal{L}^N(E \setminus K_n) \le \frac{1}{n}.$$

Let

$$N := E \setminus \bigcup_{n \in \mathbb{N} \cup \{0\}} K_n.$$

Since $K_n \subset\subset D$ and $\phi$ has the $N$-property, by (5.17) we have

$$\mathcal{L}^N(\phi(E)) = \mathcal{L}^N(\phi(\bigcup_{n \in \mathbb{N} \cup \{0\}} K_n))$$

$$= \lim_{n \to +\infty} \mathcal{L}^N(\phi(K_n))$$

$$\le \int_E |J_\phi(x)| \, dx.$$

$\square$

**Lemma 5.8** *Let $D \subset \mathbb{R}^N$ be an open set and let $\phi \in C(D)^N$ be a mapping which has the $N$-property and has a weak differential almost everywhere. Assume, in addition, that $J_\phi \in L^1_{loc}(D)$. Then for every measurable set $E \subset D$ we have*

$$\int_{\mathbb{R}^N} N(\phi, E, y) \, dy \le \int_E |J_\phi(x)| \, dx. \tag{5.18}$$

**Proof**  Recall that, by Theorem 5.5, $N(\phi, E, \cdot)$ is measurable. We may assume without loss of generality that $E$ is bounded. For each $m$ integer, let $E_1^m, \ldots, E_{k_m}^m$ be a partition of $E$ into measurable sets such that

$$\text{diam}(E_j^m) < \frac{1}{m}$$

for every $j = 1, \ldots, k_m$. Setting $N(\phi, m, \cdot) := \chi_{\phi(E_1^m)} + \ldots + \chi_{\phi(E_{k_m}^m)}$ by Theorem 5.6 we have

$$\int_{\mathbb{R}^N} N(\phi, m, y) \, dy = \sum_{j=1}^{k_m} \mathcal{L}^N(\phi(E_j^m)) \le \sum_{m=1}^{k_m} \int_{E_j^m} |J_\phi(x)| \, dx = \int_E |J_\phi(x)| \, dx.$$

$$\tag{5.19}$$

Fatou's Lemma, together with the fact that $N(\phi, m, y)$ is a nondecreasing sequence converging to $N(\phi, E, y)$ (see the first part of the proof of Theorem 5.5), yields (5.18). □

**Lemma 5.9** *Let $D \subset \mathbb{R}^N$ be an open set, let $\phi \in C(D)^N$, and let $x_0 \in D$ such that $\phi$ is differentiable at $x_0$ (in the classical sense). Assume that $J_\phi(x_0) \neq 0$. Then there is $r_0 > 0$ such that for every $0 < r \leq r_0$ the following assertions hold:*

$$\phi(x_0 + h) \neq \phi(x_0) \quad \text{for every } h \in \bar{B}(0, r) \setminus \{0\}, \tag{5.20}$$

$$d(\phi, B(x_0, r), \phi(x_0)) = \operatorname{sgn}(J_\phi(x_0)). \tag{5.21}$$

**Proof** Since $\phi$ is differentiable at $x_0$, there exists $R_0 > 0$ such that

$$\phi(x_0 + h) = \phi(x_0) + \nabla\phi(x_0)h + |h|\epsilon(|h|),$$

for $h \in \bar{B}(0, R_0)$ and $\lim_{t \to 0} \epsilon(t) = 0$. Let

$$a := \inf\{|\nabla\phi(x_0)h| : h \in \mathbb{R}^N, |h| = 1\}.$$

Since $J_\phi(x_0) \neq 0$ we obtain that $a > 0$ and we may find $0 < r_0 \leq R_0$ such that $|\epsilon(t)| \leq \frac{a}{2}$ for every $|t| \leq r_0$.

*Claim 1.* $\phi(x_0 + h) \neq \phi(x_0)$ for every $h \in \bar{B}(0, r_0)$ such that $h \neq 0$.

Indeed, if $0 < |h| \leq r_0$, then $|\epsilon(|h|)| \leq \frac{a}{2}$ and so

$$\left|\frac{\phi(x_0 + h) - \phi(x_0)}{|h|}\right| \geq \left|\nabla\phi(x_0)\frac{h}{|h|}\right| - |\epsilon(|h|)| \geq \frac{a}{2} > 0$$

and we obtain (5.20).

*Claim 2.* $d(\phi, B(x_0, r), \phi(x_0)) = \operatorname{sgn}(J_\phi(x_0))$ for every $0 < r \leq r_0$.

Let us first note that from (5.20) $d(\phi, B(x_0, r), \phi(x_0))$ is well defined for every $0 < r \leq r_0$. Fix $0 < r \leq r_0$ and set

$$u(x_0 + h) := \phi(x_0) + \nabla\phi(x_0)h, \quad h \in \bar{B}(0, r).$$

The function $u$ is an affine mapping defined on $\bar{B}(x_0, r)$. Since $J_\phi(x_0) \neq 0$ we know that $u$ is a one-to-one mapping and so $u(x_0) \neq u(x)$ for every $x \in \partial B(x_0, r)$, i.e. $\phi(x_0) \neq u(x)$ for every $x \in \partial B(x_0, r)$. Therefore, $d(u, B(x_0, r), \phi(x_0))$ is well defined and $d(u, B(x_0, r), \phi(x_0)) = \operatorname{sgn}(J_\phi(x_0))$. The application $H$ defined by

$$H(x, t) := t\phi(x) + (1 - t)u(x), \quad x \in \bar{B}(x_0, r), \ t \in [0, 1],$$

is a homotopy between $\phi$ and $u$. Moreover, for every $x \in \partial B(x_0, r)$ we have

$$|H(x, t) - \phi(x_0)| = r\left(\left|\nabla\phi(x_0)\frac{x - x_0}{|x - x_0|} + t\epsilon(|x - x_0|)\right|\right) \geq \frac{ra}{2} > 0.$$

Therefore, $\phi(x_0) \notin H(\partial B(x_0, r), t)$ for every $t \in [0, 1]$, and we conclude that $d(H(\cdot, t), B(x_0, r), \phi(x_0))$ is well defined. By virtue of Theorem 2.3, the value of $d(H(\cdot, t), B(x_0, r), \phi(x_0))$ is independent of $t$, hence, taking $t = 0, t = 1$, we obtain

$$d(\phi, B(x_0, r), \phi(x_0)) = d(u, B(x_0, r), \phi(x_0)) = \operatorname{sgn}(J_\phi(x_0))$$

and (5.21) is proved.    □

The following is a version of the latter lemma for weakly differentiable mappings.

**Lemma 5.10** *Let $D \subset \mathbb{R}^N$ be an open set and let $\phi \in C(D)^N$ be such that $\phi$ has a weak differential at $x_0 \in D$ and $J_\phi(x_0) \neq 0$. Then there exists a sequence $\{r_m : m \in \mathbb{N}\} \subset (0, 1)$ such that*

$$\lim_{m \to +\infty} d(\phi, B(x_0, h_m), \phi(x_0)) = \operatorname{sgn}(J_\phi(x_0)) \tag{5.22}$$

*and*

$$\liminf_{m \to +\infty} \frac{\mathcal{L}^N(\phi(B(x_0, r_m)))}{\mathcal{L}^N(B(x_0, r_m))} \geq |J_\phi(x_0)|. \tag{5.23}$$

**Proof**  Let $L_h : \partial B(0, 1) \to \mathbb{R}^N$ be defined by

$$L_h(y) := \frac{\phi(x_0 + hy) - \phi(x_0)}{h} - \nabla\phi(x_0)y, \quad y \in \partial B(0, 1).$$

There exists a sequence $\{h_m : m \in \mathbb{N}\} \subset (0, 1)$ converging to 0 such that

$$L_{h_m} \to 0$$

uniformly in $\partial B(0, 1)$, i.e.

$$\epsilon_m := \max_{y \in \partial B(0,1)} |L_{h_m}(y)| \underset{m \to +\infty}{\to} 0 . \tag{5.24}$$

Define $L : \mathbb{R}^N \to \mathbb{R}^N$ by

$$L(y) := \nabla\phi(x_0)(y - x_0) + \phi(x_0).$$

We observe that, for $z \in (0, 1)$, $L$ transforms the ball $B(x_0, zh_m)$ into a set of volume

$$\mathcal{L}^N(B(x_0, zh_m))|\det L| = \mathcal{L}^N(B(x_0, zh_m))|J_\phi(x_0)|. \tag{5.25}$$

Set

$$\delta := \min\{|L(y)|_2 : |y|_2 = 1\}.$$

Since $J_\phi(x_0) \neq 0$ we have $\delta > 0$ and, given $z \in (0, 1)$, we choose $m_0 \in \mathbb{N}$ such that

$$\epsilon_m < \delta(1 - z),$$

for every $m > m_0$.

*Claim 1.* We claim that

$$|y_1 - y_2|_2 > \delta(1 - z)h_m$$

for every $y_1 \in L(\partial B(x_0, h_m))$ and for every $y_2 \in L(B(x_0, zh_m))$. Indeed,

$$y_1 = \nabla\phi(x_0)(z_1 - x_0) + \phi(x_0)$$

for some $z_1$ such that $|z_1 - x_0| = h_m$ and

$$y_2 = \nabla\phi(x_0)(z_2 - x_0) + \phi(x_0)$$

for some $z_2$ such that $|z_2 - x_0| < zh_m$. We have

$$|y_1 - y_2|_2 = |\nabla\phi(x_0)(z_1 - z_2)| \geq \delta|z_1 - z_2| \geq \delta(|z_1 - x_0| - |x_0 - z_2|) > \delta(1 - z)h_m.$$

Let
$$H(x, t) := (1 - t)\phi(x) + tL(x), \quad x \in \bar{B}(x_0, h_m), t \in [0, 1].$$

*Claim 2.* $H(\partial B(x_0, h_m), t) \subset \mathbb{R}^N \setminus L(B(x_0, zh_m))$ for every $m \geq m_0(z)$ and every $t \in [0, 1]$.

Assume, on the contrary, that for some $t \in [0, 1]$, $z_2 \in B(x_0, zh_m)$ and for some $z_1 \in \partial B(x_0, h_m)$, we have $H(z_1, t) = L(z_2)$, i.e. $(1 - t)\phi(z_1) + tL(z_1) = L(z_2)$. Then
$$(1 - t)(\phi(z_1) - L(z_1)) = L(z_2) - L(z_1).$$

and by Claim 1 we have

$$\delta(1 - z)h_m < |L(z_2) - L(z_1)|_2 \leq |\phi(z_1) - L(z_1)|_2.$$

Using (5.24) and the fact that

$$z_1 = x_0 + h_m a$$

for some $a \in \partial B(0, 1)$, we deduce that

$$\delta(1 - z)h_m < \epsilon_m,$$

which yields a contradiction. Hence,

$$H(\partial B(x_0, h_m), t) \subset \mathbb{R}^N \setminus L(B(x_0, zh_m))$$

for every $t \in [0, 1]$ and every $m \geq m_0(z)$. Thus $d(H(\cdot, t), B(x_0, h_m), y)$ is well defined for every $y \in L(B(x_0, zh_m))$. Using the fact that $L$ is a bijection and $H$ is a $C$ homotopy between $\phi$ and $L$, for every $y \in L(B(x_0, zh_m))$ we have

$$d(\phi, B(x_0, h_m), y) = d(L, B(x_0, h_m), y) = sgn\, J_L(x_0) \neq 0$$

and so we obtain (5.22). Furthermore, by Theorem 2.1 we obtain that $y \in \phi(B(x_0, h_m))$ thus

$$L(B(x_0, zh_m)) \subset \phi(B(X_0, h_m))$$

for every $z \in (0, 1)$ and every $m \geq m_0(z)$, which, together with (5.25), yields

$$\liminf_{m \to +\infty} \frac{\mathcal{L}^N(\phi(B(x_0, h_m)))}{\mathcal{L}^N(B(x_0, h_m))} \geq \liminf_{m \to +\infty} \frac{\mathcal{L}^N(L(B(x_0, zh_m)))}{\mathcal{L}^N(B(x_0, h_m))} = z^N |J_\phi(x_0)|,$$

for every $z \in (0, 1)$. Letting $z$ go to 1 we conclude (5.23).     □

**Theorem 5.11** *Let $D \subset \mathbb{R}^N$ be an open, bounded set and let $\phi \in C(D)^N$ be a mapping which has a weak differential almost everywhere and has the N-property. Assume that $J_\phi \in L^1_{loc}(D)$. Then for every measurable set $E \subset D$ we have*

$$\int_E |J_\phi(x)| \, dx = \int_{\mathbb{R}^N} N(\phi, E, y) \, dy.$$

**Proof**  Recall that, by Theorem 5.5, $N(\phi, E, \cdot)$ is measurable. Also, by Lemma 5.8 it suffices to show that

$$\int_E |J_\phi(x)| \, dx \leq \int_{\mathbb{R}^N} N(\phi, E, y) \, dy.$$

Define

$$\mu(A) := \int_{\mathbb{R}^N} N(\phi, A, y) \, dy$$

for every measurable set $A \subset \mathbb{R}^N$. Then $\mu$ is positive a measure on $\mathbb{R}^N$ and since $J_\phi \in L^1_{loc}(D)$, by Lemma 5.8 we have that $\mu$ is a Radon measure. Moreover, $\mathcal{L}^N(A) = 0$ implies $\mu(A) \leq \int_A |J_\phi(a)| \, dx = 0$; therefore, by the Radon–Nikodym Theorem, there exists a measurable, positive function $\Phi$ on $D$ such that for every measurable set $A \subset D$ we have

$$\mu(A) = \int_A \Phi(x) \, dx.$$

Let $x \in D$ be such that $\phi$ has a weak differential at $x$. By Lemma 5.10 there exists a sequence $\{h_m : m \in \mathbb{N}\}$ converging to zero when $m$ tends to $+\infty$ such that

$$\liminf_{m \to +\infty} \frac{\mathcal{L}^N(\phi(B(x, h_m)))}{\mathcal{L}^N(B(x, h_m))} \geq |J_\phi(x)|.$$

Since

$$\mathcal{L}^N(\phi(B(x, h_m))) \leq \int_{\mathbb{R}^N} N(\phi, B(x, h_m), y) \, dy = \mu(B(x, h_m)),$$

we have

$$\liminf_{m \to +\infty} \frac{\mu(B(x, h_m))}{\mathcal{L}^N(B(x, h_m))} \geq |J_\phi(x)|.$$

By the Differentiation Theorem for Radon measures, we deduce that

$$\phi(x) \geq |J_\phi(x)|$$

almost everywhere in $D$ and so

$$\int_{\mathbb{R}^N} N(\phi, E, y)\, dy = \int_E \Phi(x) dx \geq \int_E |J_\phi(x)|\, dx.$$

□

## 5.2   Weakly monotone functions

In this section we study regularity properties for functions $\phi \in W^{1,p}(D)^N$ satisfying $J_\phi(x) > 0$ $\mathcal{L}^N$ a.e. $x \in D$, where $D \subset \mathbb{R}^N$ is an open set. Following Gold'sthein and Vodopyanov (1977), we show that if $p = N$, then $\phi$ is continuous and monotonic (see Theorems 5.14 and 5.17 ). The definition of monotone functions was introduced by Lebesgue (1907) and, heuristically, a function is said to be monotone if it satisfies certain weak maximum and minimum principles.

The following class of functions was introduced by Ball (1978) in his fundamental work on nonlinear elasticity and it will be used at length here,

$$\mathcal{A}_{p,q}^+(D) := \{\phi \in W^{1,p}(D)^N : \text{adj}(\nabla\phi) \in L^q(D)^{N\times N}, J_\phi > 0 \text{ a.e. } x \in D\}$$

where $\text{adj}(\nabla\phi)$ is the matrix of the cofactors of $\nabla\phi$, $p > 1$, and $q \geq \frac{p}{p-1}$. Švérak (1988) proved that if $p > N - 1$, then the mappings of $\mathcal{A}_{p,q}^+(D)$ are continuous except perhaps on a set of $p$-capacity zero and this set is empty if $p = N$ (see Theorem 5.17 and Remark 5.18). Here, to obtain Švérak's (1988) results we follow Manfredi's (1994) approach.

**Definition 5.12** *Let $D \subset \mathbb{R}^N$ be an open set, $1 \leq p < +\infty$ and let $f \in W_{loc}^{1,p}(D)$. We say that $f$ is* weakly monotone *if for every $\Omega \subset\subset D$ open, bounded, connected set and for every pair of constants $m \leq M$ such that*

$$(m - f)^+ \in W_0^{1,p}(\Omega)^N \quad \text{and} \quad (f - M)^+ \in W_0^{1,p}(\Omega)^N,$$

*we have that*

$$m \leq f(x) \leq M$$

*for almost every $x \in \Omega$.*

We recall that, if $t \in \mathbb{R}$,

$$t^+ := \begin{cases} t & t > 0 \\ 0 & t \leq 0, \end{cases} \qquad t^- := \begin{cases} 0 & t > 0 \\ -t & t \leq 0. \end{cases}$$

**Remark 5.13** *If $f \in W_{loc}^{1,p}(D) \cap C(D)$, then $f$ is weakly monotone if and only if*

$$\sup_{x\in\partial\Omega} f(x) = \sup_{x\in\partial\Omega} f(x) \quad \text{and} \quad \inf_{x\in\partial\Omega} f(x) = \inf_{x\in\partial\Omega} f(x).$$

We give a sufficient condition for a Sobolev function to be weakly monotone.

**Theorem 5.14** *Let $D \subset \mathbb{R}^N$ be an open set, $N - 1 < p < +\infty$, $q \geq \frac{p}{p-1}$ and let $\phi \in W^{1,p}(D)^N$. Assume that $J_\phi > 0$ $\mathcal{L}^N$ a.e. $x \in D$ and $\mathrm{adj}(\nabla \phi) \in L^q(D)^{N \times N}$. Then for each $i = 1, \ldots, N$ the ith component $\phi_i$ is weakly monotone. In particular, if $\phi \in W^{1,N}(D)^N$ and if $J_\phi > 0$ $\mathcal{L}^N$ a.e. $x \in D$, then $\phi$ is continuous.*

**Proof**  Fix $i = 1, \ldots, N$, let $\Omega \subset\subset D$ be an open, bounded, connected set and let $m \leq M$ be a pair of constants such that

$$(m - \phi_i)^+ \in W_0^{1,p}(\Omega)^N \quad \text{and} \quad (\phi_i - M)^+ \in W_0^{1,p}(\Omega)^N.$$

Set $g := (m - \phi_i)^+$ and let $\{\phi^r : r \in \mathbb{N}\} \subset C^\infty(D)$ be a sequence such that

$$\phi^r \to \phi \quad \text{in } W^{1,p}(\Omega)^N.$$

By Exercise 1.3 and Theorem 4.42, we have

$$\sum_{k=1}^N \frac{\partial}{\partial x_k} (\mathrm{adj}(\nabla \phi^r))_{ik} = 0 \tag{5.26}$$

and by Theorem 4.42 we obtain

$$\int_{\Omega \cap \{\phi_i < m\}} J_\phi \, dx = \int_{\Omega \cap \{\phi_i < m\}} \sum_{k=1}^N \frac{\partial \phi_i}{\partial x_k} (\mathrm{adj}(\nabla \phi))_{ik} \, dx$$

$$= \lim_{r \to +\infty} \int_{\Omega \cap \{\phi_i < m\}} \sum_{k=1}^N \frac{\partial \phi_i}{\partial x_k} (\mathrm{adj}(\nabla \phi^r))_{ik} \, dx$$

$$= \lim_{r \to +\infty} \int_\Omega \sum_{k=1}^N \chi_{\{\phi_i < m\}} \frac{\partial \phi_i}{\partial x_k} (\mathrm{adj}(\nabla \phi^r))_{ik} \, dx$$

$$= - \lim_{r \to +\infty} \int_\Omega \sum_{k=1}^N \frac{\partial g}{\partial x_k} (\mathrm{adj}(\nabla \phi^r))_{ik} \, dx$$

$$= \lim_{r \to +\infty} \int_\Omega \sum_{k=1}^N g \frac{\partial}{\partial x_k} (\mathrm{adj}(\nabla \phi^r))_{ik} \, dx$$

$$= 0.$$

Since $J_\phi(x) > 0$ $\mathcal{L}^N$ a.e. $x \in \Omega$, we deduce that

$$\mathcal{L}^N(\{\phi_i < m\}) = 0$$

and so

$$\phi_i(x) \geq m \quad \text{a.e. } x \in \Omega.$$

Using a similar argument we obtain that

$$\phi_i(x) \leq M \quad \text{a.e. } x \in \Omega,$$

thus $\phi_i$ is weakly monotone. □

Recall that if $X \subset \mathbb{R}^N$ is a $C^\infty$ paracompact manifold of dimension $r \in \mathbb{N}$, then $X$ can be covered by finitely many manifolds $X_i \subset X$ such that for each $i$ there exists a $C^\infty$ diffeomorphism $u_i : (0,1)^r \to X_i$. We say that $f \in W^{1,p}(X)$ if and only if $f \circ u_i \in W^{1,p}((0,1)^r)$. Recall also that, given any ball $B(x_0, r) \subset \mathbb{R}^N$, $\partial B(x_0, r)$ is a $C^\infty$ paracompact manifold of dimension $N - 1$.

**Theorem 5.15** *Let $N - 1 < p \leq N$ and let $f \in W^{1,p}(\partial B(x_0, r))$ be a continuous function on the sphere $\partial B(x_0, r)$. Then,*

$$(\text{diam}(f(\partial B(x_0, r))))^p \leq C(N, p) r^{p+1-N} \int_{\partial B(x_0, r)} |\nabla f(x)|_2^p \, dH^{N-1},$$

*where $C(N, p)$ is a constant depending only on $N$ and $p$.*

**Proof** Since $p > \text{diam}(\partial B(x_0, r)) = N - 1$, by the Sobolev Imbedding Theorem expressed on $\partial B(x_0, r)$, we have

$$(\text{diam}(f(B(x_0, r))))^p \leq C(r, p, N) \int_{\partial B(x_0, r)} |\nabla f(x)|_2^p \, dH^{N-1}.$$

Using a rescaling argument, we conclude that

$$C(r, p, N) = r^{p+1-N} C(N, p).$$

□

In the sequel, given $f \in W^{1,p}(D)$ for some open set $D \subset \mathbb{R}^N$, we denote by $f^*$ the $p$-quasicontinuous representative of $f$ of Theorem 4.61 and by $T_r : W^{1,p}(B(x_0, R)) \to L^p(\partial B(x_0, r))$ the trace operator for every $r$ such that $B(x_0, r) \subset\subset D$ (see Theorem 4.50).

**Theorem 5.16** *Let $D \subset \mathbb{R}^N$ be an open, bounded set, $1 \leq p \leq +\infty$, $f \in W^{1,p}(D)$. Then*

(i) *$f^*|_{\partial B(x_0, r)} \in W^{1,p}(\partial B(x_0, r))$ for $H^1$ almost every $r$ such that $B(x_0, r) \subset\subset D$;*

(ii) *$f^*(x) = T_r(x)$ for $H^{N-1}$ almost every $x \in \partial B(x_0, r)$, for every $r$ such that $B(x_0, r) \subset\subset D$;*

(iii) *if $p > N - 1$ and if $B(x_0, h) \subset\subset D$, then $f$ admits a representative $\bar{f} \in W^{1,p}(B(x_0, h))$ such that $\bar{f}|_{\partial B(x_0, r)}$ is continuous on $\partial B(x_0, r)$ for $H^1$ almost every $r \in (0, h)$.*

**Proof** We assume, without loss of generality, that $f^* \equiv f$.

(i) Let $x_0 \in D$ and let $h > 0$ be such that $B(x_0, h) \subset\subset D$. Since $f \in W^{1,p}(D)$ there exists a sequence $\{f_n : n \in \mathbb{N}\} \subset C^\infty(D)$ such that

$$\int_D |f_n - f|_2^p \, dx \leq \frac{1}{2^n}, \quad \int_D |\nabla f_n - \nabla f|_2^p \, dx \leq \frac{1}{2^n}$$

for every $n \in \mathbb{N}$. Set

$$F_n(r) := \int_{\partial B(x_0, r)} |f_n - f|_2^p \, dH^{N-1},$$

$$G_n(r) := \int_{\partial B(x_0, r)} |\nabla f_n - \nabla f|_2^p \, dH^{N-1}.$$

We observe that

$$\int_0^h F_n(r) \, dr, \ \int_0^h G_n(r) \, dr \leq \frac{1}{2^n}$$

for every $n \in \mathbb{N}$ and so

$$\sum_{n=1}^\infty \int_0^h F_n(r) \, dr = \sum_{n=1}^\infty \int_{B(x_0, h)} |f_n - f|_2^p \, dx \leq \sum_{n=1}^\infty \frac{1}{2^n} < +\infty,$$

$$\sum_{n=1}^\infty \int_0^h G_n(r) \, dr = \sum_{n=1}^\infty \int_{B(x_0, h)} |\nabla f_n - \nabla f|_2^p \, dx \leq \sum_{n=1}^\infty \frac{1}{2^n} < +\infty.$$

Since $F_n(r), G_n(r) \geq 0$, by the Lebesgue Dominated Convergence Theorem we obtain that

$$\sum_{n=1}^\infty F_n \in L^1((0, h)), \ \sum_{n=1}^\infty G_n \in L^1((0, h))$$

and so there exists $I \subset (0, h)$ such that $\mathcal{L}^1(I) = 0$ and

$$\sum_{n=1}^\infty F_n(r) < +\infty, \ \sum_{n=1}^\infty G_n(r) < +\infty,$$

for every $r \in (0, h) \setminus I$. Thus,

$$\lim_{n \to +\infty} F_n(r) = 0 \text{ and } \lim_{n \to +\infty} G_n(r) = 0 \tag{5.27}$$

for every $r \in (0, h) \setminus I$ and we conclude that

$$f|_{\partial B(x_0, r)} \in W^{1,p}(\partial B(x_0, r))$$

for every $r \in (0, h) \setminus I$.

(ii) By (i) for $\mathcal{L}^1$ almost every $r$ such that $B(x_0, r) \subset\subset D$ we have $f|_{\partial B(x_0,r)} \in W^{1,p}(\partial B(x_0, r))$ and, for each such $r$, by (5.27), there exists a subsequence such that $f_{n_k}(x) \to f(x)$ for $H^{N-1}$ a.e. $x \in \partial B(x_0, r)$ as $k \to \infty$. On the other hand, $T_r(f_{n_k}) = f_{n_k}|_{\partial B(x_0,r)}$ and $T_r(f_{n_k})(x) \to T_r(f)(x)$ for $H^{N-1}$ a.e. $x \in \partial B(x_0, r)$ as $k \to \infty$. We conclude that $T_r(f)(x) = f(x)$ for $H^{N-1}$ a.e. $x \in \partial B(x_0, r)$.

(iii) Let $h > 0$ be such that that $B(x_0, h) \subset\subset D$. By assertions (i) and (ii) and the Sobolev Imbedding Theorem on $\partial B(x_0, r)$, there exists $J \subset (0, h)$ such that $\mathcal{L}^1(J) = 0$ and $T_r(f)$ has a continuous representative $g_r$ on $\partial B(x_0, r)$ for every $r \in (0, h) \setminus J$. Define $\bar{f}$ by

$$\bar{f}(x) := \begin{cases} g_r(x) & \text{if } |x - x_0| = r \in (0, h) \setminus J \\ f^*(x) & \text{if } |x - x_0| = r \in J. \end{cases}$$

We observe that

$$\bar{f}(x) = f^*(x)$$

for $\mathcal{L}^N$ almost every $x \in B(x_0, h)$ and $\bar{f}$ has the property required in assertion (iii).

$\square$

**Theorem 5.17** *Let $D \subset \mathbb{R}^N$ be an open set, $N - 1 < p \leq N$, and let $f \in W^{1,p}(D)$ be a weakly monotone function. Then $f$ admits a representative (still denoted by $f$) such that $f|_{D \setminus S} : D \setminus S \to \mathbb{R}$ is continuous for some set $S \subset D$ such that*

$$Cap_{p-a}(S) = 0$$

*for every $0 < a < p$. If $p = N$, then $S = \emptyset$ and $f$ is continuous in $D$. In particular, if $f \in W^{1,N}(D)^N$ and if $J_f > 0$ $\mathcal{L}^N$ a.e. in $D$, then $f$ is continuous in $D$.*

The argument we use in the proof below was introduced by Gold'sthein and Reshetnyak (1990) in the case where $p \geq N$, and later extended by Manfredi (1994) to the case where $p > N - 1$. In fact, in the theorem above we can choose the set $S$ such that its Hausdorff dimension is equal to $N - p$.

**Proof** Let $x_0 \in D$, let $0 < h < 2$ be such that $B(x_0, h) \subset\subset D$, and assume, without loss of generality, that

$$f \equiv \bar{f},$$

where $\bar{f}$ is the representative of $f$ given by Theorem 5.16 (iii). Let $J \subset (0, h)$ be such that $\mathcal{L}^1(J) = 0$, $\bar{f}|_{\partial B(x_0,r)}$ is continuous and

$$\bar{f}|_{\partial B(x_0,r)} \in W^{1,p}(\partial B(x_0, r))$$

for every $r \in (0, h) \setminus J$. Set

$$m_r(x_0) := \inf\{f(x) : x \in \partial B(x_0, r)\}$$

and

$$M_r(x_0) := \sup\{f(x) : x \in \partial B(x_0, r)\},$$

for every $r \in (0, h)$.

*Claim 1.* $m_r(x_0) \leq f(x) \leq M_r(x_0)$ for $\mathcal{L}^N$ almost every $x \in B(x_0, r)$ and for every $r \in (0, h) \setminus J$.

Indeed, for every $r \in (0, h) \setminus J$

$$T_r(m_r(x_0) - f)^+ \equiv 0$$

and

$$T_r(f - M_r(x_0))^+ \equiv 0$$

and since $f$ is weakly monotone on $D$, we deduce that

$$m_r(x_0) \leq f(x) \leq M_r(x_0)$$

for $\mathcal{L}^N$ almost every $x \in B(x_0, r)$ and for every $r \in (0, h) \setminus J$. Define

$$\text{ess osc}_{B(x_0, r)} = \text{ess sup}\{f(x) : x \in B(x_0, r)\} - \text{ess inf}\{f(x) : x \in B(x_0, r)\}$$
$$=: c(r, x_0)$$

and

$$d(r) := \int_r^h t^{p-N} dt \int_{\partial B(x_0, r)} |\nabla f|_2^p \, dH^{N-1}.$$

Let $C(N, p)$ be the constant in Theorem 5.15.

*Claim 2.* $(c(r, x_0))^p \log \frac{h}{r} \leq C(N, p) d(r)$ for every $r \in (0, h)$.

Indeed, by Theorem 5.15 and the fact that $f \equiv \bar{f}$,

$$(M_t(x_0) - m_t(x_0))^p \leq C(N, p) t^{p-N+1} dt \int_{\partial B(x_0, t)} |\nabla f|_2^p \, dH^{N-1}, \tag{5.28}$$

for every $t \in (0, h) \setminus J$. By Claim 1 we have

$$M_t(x_0) - m_t(x_0) = c(t, x_0)$$

for every $t \in (0, h) \setminus J$ and so,

$$c(t, x_0)^p \leq C(N, p) t^{p-N+1} dt \int_{\partial B(x_0, t)} |\nabla f|_2^p \, dH^{N-1} \tag{5.29}$$

for every $t \in (0, h) \setminus J$. Dividing both sides of (5.29) by $t$ and integrating from $r$ to $h$ we obtain

$$\int_r^h \frac{(c(t,x_0))^p}{t}\, dt \le C(N,p)\, d(r).$$

Since $c$ is nondecreasing on $(0,h)$, we deduce that

$$c(r,x_0)^p \log \frac{h}{r} \le C(N,p) d(r)$$

and so if, moreover, $0 < r < 1$, setting $h = \sqrt{r}$, we have

$$\left(\mathrm{ess\,osc}_{B(x_0,r)}\right)^p \le \frac{C(N,p)}{\log \frac{1}{\sqrt{r}}} \int_{B(x_0,r)} \frac{|\nabla f(y)|_2^p}{|y - x_0|_2^{N-p}}\, dy \tag{5.30}$$

for every $r \in (0,h)$. Define $w : \mathbb{R}^N \to \mathbb{R}$ by

$$w(x) := \begin{cases} |\nabla f(x)|_2^p & x \in D \\ 0 & x \notin D. \end{cases}$$

Let

$$I_p(x) = |x|_2^{p-N}, \quad x \in \mathbb{R}^N,\ N-1 < p < N$$

be the Riesz kernel of order $p$ (see Stein 1970),

$$S := \{x \in D : I_p * w(x) = +\infty\}.$$

We observe that

$$S = \emptyset \quad \text{if } p = N.$$

By (5.30) we obtain

$$\lim_{r \to 0+} c(r,x) = 0 \tag{5.31}$$

for every $x \in D \setminus S$ and by Theorem 4.65 $Cap_{p-a}(S) = 0$ for every $0 < a < p$. Indeed, since $w \ge 0$, $I_p * w$ is lower semicontinuous and so $S$ is a Borel set. This implies that (see Hayman and Kennedy 1976, Theorem 5.3)

$$C_{p-a}(S) = \sup\{C_{p-a}(K) : K \subset S, K \text{ compact set}\},$$

and, by Theorem 4.65, we obtain $C_{p-a}(K) = 0$ for every compact set $K \subset S$. By Theorem 4.61 we may find a set $E \subset D$ such that $Cap_p(E) = 0$ and

$$f(x) = \lim_{r \to 0+} \frac{1}{\mathcal{L}^N(B(x,r))} \int_{B(x,r)} f(y)\, dy \tag{5.32}$$

for every $x \in D \setminus E$.

*Claim 3.* $f$ is continuous on $D \setminus (S \cup E)$.

Let $x \in D \, (S \cup E)$ and let $\{x_n \; : \; n \in \mathbb{N}\} \subset D \setminus S$ be a sequence converging to $x$. Let $\epsilon, \delta > 0$ be such that

$$\left[ \frac{C(N, p)}{\log \frac{1}{\sqrt{2\delta}}} \left( \int_{B(x_0, 2\delta)} \frac{|\nabla f(y)|_2^p}{|y - x|_2^{N-p}} \, dy \right) \right]^{\frac{1}{p}} < \frac{\epsilon}{2 \mathcal{L}^N(B(0, 1))}. \tag{5.33}$$

Assume that $|x_n - x|_2 < \delta$. By (5.32) there exists $r(x) > 0$ such that

$$-\frac{\epsilon}{4} < f(x) - \frac{1}{\mathcal{L}^N(B(x, r))} \int_{B(x, r)} f(y) \, dy < \frac{\epsilon}{4}$$

for every $0 < r < r(x)$ and there exists $0 < r(x_n) \leq r(x)$ such that

$$-\frac{\epsilon}{4} < f(x_n) - \frac{1}{\mathcal{L}^N(B(x_n, r))} \int_{B(x_n, r)} f(y) \, dy < \frac{\epsilon}{4}$$

for every $0 < r < r(x_n)$ and every $n \in \mathbb{N}$. Therefore,

$$-\frac{\epsilon}{2} < f(x) - f(x_n) + \frac{1}{\mathcal{L}^N(B(x_n, r))} \int_{B(x_n, r)} f(y) \, dy$$
$$- \frac{1}{\mathcal{L}^N(B(x, r))} \int_{B(x, r)} f(y) \, dy$$
$$< \frac{\epsilon}{2}$$

for every $0 < r < r(x_n)$ and every $n \in \mathbb{N}$ and so

$$-\frac{\epsilon}{2} < f(x) - f(x_n) + \frac{1}{\mathcal{L}^N(B(0, 1))} \int_{B(0,1)} r^N (f(x_n + rt) - f(x + rt)) \, dt < \frac{\epsilon}{2} \tag{5.34}$$

for every $0 < r < r(x_n)$ and every $n \in \mathbb{N}$. Fix $r < \delta$. We have

$$x + rt, \; x_n + rt \in B(x, 2\delta) \quad \text{for all } t \in B(0, 1)$$

and so by (5.30) and (5.33) we have

$$|f(x_n + rt) - f(x + rt)| < \frac{\epsilon}{2 \mathcal{L}^N(B(0, 1))} \quad \text{a.e. } t \in B(0, 1).$$

Integrating in $B(0, 1)$, we obtain

$$\left| \int_{B(0,1)} r^N (f(x_n + rt) - f(x + rt)) \, dt \right| < \frac{\epsilon}{2},$$

which, together with (5.34), yields

$$|f(x_n) - f(x)| < \epsilon.$$

Thus, $f$ is continuous on $D \setminus (S \cup E)$ and, by Theorem 4.57, $Cap_{p-a}(S \cup E) = 0$.

Finally, if $f \in W^{1,N}(D)^N$ and if $J_f(x) > 0$ $\mathcal{L}^N$ a.e. $x \in D$, then, by Theorem 5.14 the components $f_i$ are weakly monotone and as $S = \emptyset$ we conclude that $f$ is continuous in $D$.      $\square$

From now on, we identify a function $\phi \in W^{1,N}(D)^N$ such that $J_\phi(x) > 0$ $\mathcal{L}^N$ a.e. $x \in D$ with its continuous representative.

**Remark 5.18** It can be shown that if $f \in W^{1,p}(D)^N$, $p > N - 1$, adj $\nabla f \in L^q$, $q \geq \frac{N}{N-1}$ and if $J_f(x) > 0$ $\mathcal{L}^N$ a.e. $x \in D$, then $f$ is continuous outside a set $S$ of Hausdorff dimension $N - p$. Also, for every $\epsilon > 0$ the set $\{x \in D : \limsup_{r \to 0+} \operatorname{osc}_{B(x,r)} < \epsilon\}$ is open. These results were stated by Müller *et al.* (1994) and their proof can be obtained exactly as in Šverák (1988), where we assume that $q \geq \frac{p}{p-1}$. We note that for $N - 1 < p < N$ we have $\frac{p}{p-1} > \frac{N}{N-1}$. Also, by Theorem 4.57, $H_{dim}(S) \leq N - p$ implies that $Cap_{p-a}(S) = 0$ for all $0 < a < p$ and we recover Theorem 4.57.

The following result is due to Reshetnyak (1989).

**Corollary 5.19** *Let $D \subset \mathbb{R}^N$ be an open set, let $K$ be a compact set, and let $V$ be an open, bounded set such that $K \subset V \subset\subset D$. Let $\phi \in W^{1,N}_{loc}(D)^N$ be such that $J_\phi(x) > 0$ $\mathcal{L}^N$ a.e. $x \in D$. Then there exists a constant $\delta > 0$ such that*

$$|\phi(x_1) - \phi(x_2)|_2 \leq C(N) M^{\frac{1}{N}} \theta(|x_1 - x_2|_2)$$

*for every $x_1, x_2 \in K$ such that $|x_1 - x_2|_2 \leq \delta$, where*

$$M := \int_V |\nabla v(x)|^N \, dx, \quad \theta(t) := \left(\frac{2}{\log(\frac{2}{t})}\right)^{\frac{1}{N}}, \quad \lim_{t \to 0+} \theta(t) = 0$$

*and $C(N)$ is the constant of Theorem 5.15.*

**Proof** Fix $i \in \{1, \dots, N\}$ and set $f := \phi_i$. Let

$$d := \operatorname{dist}(K, \mathbb{R}^N \setminus V)$$

and

$$\delta := \max \left\{ t : 0 < t \leq 2, \frac{t}{2} + \sqrt{\frac{t}{2}} \leq d \right\}.$$

Fix $x_1, x_2 \in K$ such that $|x_1 - x_2|_2 < \delta$ and set

$$h := \frac{|x_1 - x_2|_2}{2}, \quad x_0 := \frac{x_1 + x_2}{2}.$$

We claim that $B(x_0, \sqrt{h}) \subset D$. Indeed, since $h < \frac{\delta}{2}$ we have, for every $x \in B(x_0, \sqrt{h})$,

$$|x - x_1|_2 \leq |x - x_0|_2 + |x_0 - x_1|_2 \leq \sqrt{h} + h < d$$

and so

$$B(x_0, \sqrt{h}) \subset D.$$

According to Theorem 5.16 let $I \subset (0, \sqrt{h})$ be such that $\mathcal{L}^1(I) = 0$ and

$$f|_{\partial B(x_0, r)} \in W^{1, N}(\partial B(x_0, r))$$

for every $r \in (0, \sqrt{h}) \setminus I$. By Theorem 5.17 $f$ is weakly monotone and continuous and Theorem 5.15 yields

$$\left( \mathrm{diam}(f(\partial B(x_0, r))) \right)^N \leq rC(N) \int_{\partial B(x_0, r)} |\nabla f(x)|_2^N \, dH^{N-1}$$

for every $r \in (0, \sqrt{h}) \setminus I$ and for some constant $C(N)$ depending only on $N$. Again using the fact that $f$ is weakly monotone, we obtain that

$$\mathrm{diam}(f(\partial B(x_0, r))) = \mathrm{diam}(f(B(x_0, r)))$$

for every $r \in (0, \sqrt{h})$ and so

$$(\mathrm{diam}(f(B(x_0, r))))^N \leq rC(N) \int_{\partial B(x_0, r)} |\nabla f(x)|_2^N \, dH^{N-1} \tag{5.35}$$

for every $r \in (0, \sqrt{h}) \setminus I$. Dividing both sides of the inequality by $r$ and integrating with respect to $r$, we obtain

$$\int_h^{\sqrt{h}} \frac{[\mathrm{diam}(f(B(x_0, r)))]^N}{r} \, dr \leq C^N(N) \int_h^{\sqrt{h}} dr \int_{\partial B(x_0, r)} |\nabla f|_2^N \, dH^{N-1}$$

$$\leq C^N(N) \int_{B(x_0, \sqrt{h})} |\nabla f|_2^N \, dx$$

$$\leq C^N(N) M.$$

Let

$$\gamma(r) := \mathrm{diam}(f(B(x_0, r))).$$

We observe that $\gamma$ is nondecreasing in $(0, \sqrt{h})$ and so

$$\gamma(h)^N \log \frac{1}{\sqrt{h}} \leq C^N(N) M$$

which yields

$$|f(x_1) - f(x_2)| \leq C M^{\frac{1}{N}} \theta(|x_1 - x_2|_2).$$

$\square$

For Sobolev functions which do not satisfy the regularity assumptions of Corollary 5.19, we can, none the less, assert the following corollary.

**Corollary 5.20** *Let $D \subset \mathbb{R}^N$ be an open set, let $N-1 < p < N$, and let $q \geq \frac{p}{p-1}$. Assume that $\phi \in W_{loc}^{1,p}(D)^N$ is such that $\mathrm{adj}(\nabla \phi) \in L^q(D)^{N \times N}$ and $J_\phi(x) > 0$ $\mathcal{L}^N$ a.e. $x \in D$. Assume that $\phi$ coincides with its representative of Theorem 5.17. Then for every $x_0 \in D$ and for every $r > 0$ such that $B(x_0, 2r) \subset D$, we have*

$$|\phi(x_1) - \phi(x_2)|_2^p \leq \frac{C(N,p)r^p}{\mathcal{L}^N(B(x_0,r))} \int_{B(x_0,2r)} |\nabla \phi|_2^p \, dx$$

*for almost every $x_1, x_2 \in B(x_0, r)$, where $C(N, p)$ is a constant depending only on $N$ and $p$.*

**Proof** This follows from integrating (5.29) for $t \in [r, 2r]$ and using the fact that $c(\cdot, x_0)$ is nondecreasing (see the proof of Theorem 5.17). □

**Theorem 5.21** *Let $D \subset \mathbb{R}^N$ be an open set and let $\phi \in W^{1,p}(D)^N$, $p > N - 1$. Then $\phi$ has a weak differential almost everywhere. Furthermore, if $p = N$ and if the $J_\phi > 0$ $\mathcal{L}^N$ a.e. or if $p > N$, then $\phi$ has a differential almost everywhere (in the classical sense).*

**Proof** Let $x \in D$ be a Lebesgue point for $\nabla \phi$ and

$$\lim_{h \to 0} \|R_{h,x}\phi_i\|_{1,p}(B(0,1)) = 0 \tag{5.36}$$

for each $i = 1, \ldots, N$, where

$$R_{h,x}\phi_i(y) = \frac{\phi_i(x + hy) - \phi_i(x)(x)}{h} - \sum_{j=1}^{N} \frac{\partial \phi_i}{\partial x_j}(x)y_j.$$

By Theorem 4.49, (5.36) holds for $\mathcal{L}^N$ almost every $x \in D$. Set

$$L_h(y) = \frac{\phi(x + hy) - \phi(x)}{h}, \quad y \in B(0,1), 0 < |h| < \mathrm{dist}(x, \partial D)$$

and

$$L(y)_i = \sum_{j=1}^{N} \frac{\partial \phi_i}{\partial x_j}(x)y_j.$$

*Case 1. $p > N - 1$.*

   By (5.36) $\lim_{h \to 0} \|L_h - L\|_p(B(0,1)) = 0$ and as

$$\mathcal{L}^N \left( \{y \in B(0,1) : |L_h(y) - L(y)|_2 \geq \epsilon\} \right) \leq \frac{1}{\epsilon^p} \int_{B(0,1)} |L_h(y) - L(y)|_2^p \, dy$$

then $L_h$ converges to $L$ in measure. Let $\{h_n : n \in \mathbb{N}\} \subset (0,1)$ be a decreasing sequence converging to 0. By Theorem 5.16 (iii), there exists $I \subset (0,1)$ such that $\mathcal{L}^1(I) = 0$ and both $L$ and $L_{h_n}$ are continuous on $\partial B(0,r)$ for every $r \in (0,1) \setminus I$ and for all $n \in \mathbb{N}$. Applying Theorems 5.16 and 5.15 to $L_{h_n} - L$, as $(L_{h_n} - L)(0) = 0$ we deduce that

$$|L_{h_n}(y) - L(y)|^p \le C(N,p)r^{p-N+1} \int_{\partial B(0,r)} |\nabla\phi(x + h_n z) - \nabla\phi(x)|^p \, dH^{N-1}(z)$$

$$(5.37)$$

for every $y \in B(0,r)$ and for every $r \in (0,1) \setminus I$. Since $x$ is a Lebesgue point for $\nabla\phi$, we have

$$0 = \lim_{n \to +\infty} \int_{B(0,1)} |\nabla\phi(x + h_n z) - \nabla\phi(x)|_2^p \, dz$$

$$= \lim_{n \to +\infty} \int_0^1 dr \int_{\partial B(0,r)} |\nabla\phi(x + h_n z) - \nabla\phi(x)|_2^p \, dH^{N-1}(z).$$

Hence, there exists a subsequence $\{n_k\}$ such that

$$\int_{\partial B(0,r)} |\nabla\phi(x + h_{n_k} z) - \nabla\phi(x)|_2^p \, dH^{N-1}(z) \to 0 \qquad (5.38)$$

as $k \to \infty$, for $\mathcal{L}^1$ a.e. $r \in (0,1)$, and, in particular, we may fix $r \in (0,1) \setminus I$ for which (5.38) holds. Finally, (5.37), together with (5.38), yields

$$\lim_{k \to \infty} \operatorname{ess\,sup}\{|L_{h_{n_k}}(y) - L(y)|_2 : y \in \partial B(0,1)\} = 0$$

and so $\phi$ has a weak differential at $x$.

Case 2. $p = N$ and $J_\phi > 0$ $\mathcal{L}^N$ a.e.

We observe that

$$\det(\nabla L_h(y)) = \det(\nabla\phi(x + hy)) > 0$$

for every $0 < |h| < c := \frac{1}{2}\operatorname{dist}(x, \partial D)$ and for almost every $y \in B(0,1)$. Applying Lemma 5.19, with $D = B(0,2)$, $V = B(0, \frac{3}{2})$, and $K = \bar{B}(0,1)$, we obtain that $\{L_h : 0 < |h| < c\}$ is equicontinuous on $K$. By the Ascoli–Arzelà Theorem $L_h$ converges uniformly to some $G \in C(K, \mathbb{R}^N)$ and since, by (5.36), $L_h$ converges to $L$ in the $L^N(B(0,1), \mathbb{R}^N)$ norm, we conclude that $G = L$. Hence

$$\lim_{h \to 0} \sup\{|L_h(y) - L(y)|_2 : y \in \bar{B}(0,1)\} = 0$$

and so $\phi$ has a differential in the classical sense at $x$.

Case 3. $p > N$.

By Sobolev's Imbedding Theorem and (5.36), we obtain that $R_{h,x}$ converges to 0 in $C^{0,\alpha}(D)^N$, where $\alpha = 1 - \frac{N}{p}$. Thus,

$$\lim_{h \to 0} \sup\{|L_h(y) - L(y)|_2 : y \in \bar{B}(0,1)\} = 0$$

and so $\phi$ has a differential in the classical sense at $x$.    □

## 5.3    Change of variables via the multiplicity function

**Proposition 5.22** *Let $D \subset \mathbb{R}^N$ be an open, bounded set and assume that $\phi \in C(D)^N$ has a weak differential almost everywhere; it has the N-property and $J_\phi \in L^1_{loc}(D)$. Let $v : \mathbb{R}^N \to \mathbb{R}$ be a measurable function and define $u : \mathbb{R}^N \to \mathbb{R}$ by*

$$u(x) := \begin{cases} v \circ \phi J_\phi(x) & \text{if } J_\phi(x) \text{ exists} \\ 0 & \text{otherwise.} \end{cases}$$

*Then $u$ is a measurable function and $\phi^{-1}(E)$ is measurable for every measurable set $E \subset \mathbb{R}^N$. Furthermore, $\mathcal{L}^N(\phi^{-1}(B) \setminus A) = 0$ whenever $B$ is a measurable set, $\mathcal{L}^N(B) = 0$, and $A := \{x \in D : J_\phi(x) = 0\}$.*

Note that we did not assert that $v \circ \phi$ is measurable, since we do not know if $v \circ \phi$ is well defined. Indeed, if $\phi$ does not have the $N^{-1}$-property, we cannot deduce that $v \circ \phi = w \circ \phi$ $\mathcal{L}^N$ a.e. $x \in D$ whenever $v(y) = w(y)$ for $\mathcal{L}^N$ a.e. $y \in \mathbb{R}^N$.

**Proof of Proposition 5.22**
*Step 1.* We prove that $u$ is well defined, i.e. $u(x) = \bar{u}(x)$ $\mathcal{L}^N$ a.e. $x \in D$, where $\bar{u}$ is defined by

$$\bar{u}(x) := \begin{cases} w \circ \phi J_\phi(x) & \text{if } J_\phi(x) \text{ exists} \\ 0 & \text{otherwise.} \end{cases}$$

and $w : \mathbb{R}^N \to \mathbb{R}$ is a measurable function such that $v(y) = w(y)$ for every $y \in \mathbb{R}^N \setminus L$ where $L \subset \mathbb{R}^N$ is such that $\mathcal{L}^N(L) = 0$. Let $E \subset \mathbb{R}^N$ be a measurable set, let

$$A := \{x \in D : J_\phi(x) = 0\},$$

and let $P, Q \subset \mathbb{R}^N$ be Borel measurable sets such that

$$P \subset E \subset Q, \quad \mathcal{L}^N(Q \setminus P) = 0.$$

As $\phi$ is continuous, $\phi^{-1}(P)$ and $\phi^{-1}(Q \setminus P)$ are Borel measurable and, by Theorem 5.11, we obtain

$$\int_{\phi^{-1}(Q \setminus P)} |J_\phi(x)| \, dx \leq \int_{Q \setminus P} N(\phi, \phi^{-1}(Q \setminus P), y) \, dy = 0.$$

Thus,

$$\mathcal{L}^N(\phi^{-1}(Q \setminus P) \setminus A)) = 0,$$

which, together with the fact that $\phi^{-1}(P)$ is Borel measurable and

$$\phi^{-1}(P) \setminus A \subset \phi^{-1}(E) \setminus A \subset \phi^{-1}(P) \cup [\phi^{-1}(Q \setminus P) \setminus A],$$

yields that $\phi^{-1}(E) \setminus A$ is measurable and as $A$ is measurable so is $\phi^{-1}(E)$. In particular, if $B$ is a measurable set and $\mathcal{L}^N(B) = 0$, by Theorem 5.11 we have

$$\mathcal{L}^N(\phi^{-1}(B) \setminus A) = 0. \tag{5.39}$$

Finally, we have

$$v \circ \phi(x) = w \circ \phi(x)$$

for every $x \in D \setminus \phi^{-1}(L)$ and so

$$u(x) = \bar{u}(x)$$

$\mathcal{L}^N$ a.e. $x \in D \setminus \phi^{-1}(L)$. Also,

$$u(x) = 0 = \bar{u}(x)$$

for every $x \in A \cap \phi^{-1}(L)$ and since, by (5.39), $\mathcal{L}^N(\phi^{-1}(L) \setminus A) = 0$, we deduce that

$$u(x) = \bar{u}(x)$$

$\mathcal{L}^N$ a.e. $x \in D$ and $u$ is well defined.

*Step 2.* $u$ is measurable.

Assume without loss of generality that $u \geq 0$. For $t > 0$ set

$$E_t := \{x \in D : u(x) < t\}.$$

and

$$M := \{x \in D : J_\phi(x) \text{ does not exist}\}.$$

We have $\mathcal{L}^N(M) = 0$ and so there exists a subset $B \subset D$ with measure zero such that

$$E_t = \{x \in D \setminus (A \cup M) : u(x) < t\} \cup A \cup B$$

if $t > 0$ and $E_t = \{x \in D \setminus (A \cup M) : u(x) < t\} \cup B$ if $t \leq 0$. We observe that

$$\{x \in D \setminus (A \cup M) : u(x) < t\}$$
$$= \bigcup_{\substack{r,s \in \mathbb{Q} \\ rs < t}} [\phi^{-1}(v^{-1}(-\infty, r)) \setminus (A \cup M)] \cap \{x \in D \setminus M : J_\phi(x) < s\}$$

and by Step 1, $\phi^{-1}(v^{-1}(-\infty, r)) \setminus (A \cup M)$ is measurable for each $r \in \mathbb{Q}$. Since we already know that

$$\{x \in D \setminus M : J_\phi(x) < s\}$$

is measurable for each $s \in \mathbb{Q}$, we deduce that $\{x \in D \setminus (A \cup M) : u(x) < t\}$ is measurable, thus $u$ is measurable.                    □

We now state a change of variables formula for the multiplicity function.

**Theorem 5.23** *Let $D \subset \mathbb{R}^N$ be an open, bounded set and assume that $\phi \in C(D)^N$ has a weak differential almost everywhere, it has the N-property, and $J_\phi \in L^1_{loc}(D)$. If $v \in L^\infty(\mathbb{R}^N)$, then for every measurable set $E \subset D$,*

$$\int_E v \circ \phi(x) |J_\phi(x)| \, dx = \int_{\mathbb{R}^N} v(y) N(\phi, E, y) \, dy. \tag{5.40}$$

**Remark 5.24** For the proof of Theorem 5.23 we follow Gold'sthein and Reshet-nyak (1990). In the case where $N = 2$ the result was first obtained by Rado and Reichelderfer (1955) for $\phi \in W^{1,p}(D)^N$, $p > N$. This was extended later by Marcus and Mizel (1973) to functions $\phi \in W^{1,p}(D)^m$, $p > N$, and $N \leq m$, and it was shown that, for every $v : \mathbb{R}^m \to \mathbb{R}$ measurable and for every measurable set $E \subset D$,

$$\int_E v \circ \phi(x) J_N(\phi(x)) \, d\mathcal{L}^N(x) = \int_{\phi(E)} v(y) N(\phi, E, y) \, d\mathcal{L}^m(y)$$

whenever one of the two sides is meaningful. Here we use the notation

$$[J_N(\phi(x))]^2 = \sum_{i=1}^{N} \sum_{j=1}^{P} (F_{ij}(x))^2$$

where $F$ is the matrix of the $N \times N$ minors of $\nabla \phi(x)$ and $P := \frac{m!}{(m-N)! N!}$.

As it turns out, Theorem 5.23 will generalize this result for $m = N$, since, by Theorems 5.21 and 5.28, if $\phi \in W^{1,p}(D)^N$ and $p > N$, then $\phi$ is weakly differentiable and has the $N$-property.

**Proof of Theorem 5.23** Without loss of generality, we may assume that

$$v(y) \geq 0$$

for almost every $y \in \mathbb{R}^N$. We divide the proof into three cases.

*Case 1.* Assume that $v = \chi_F$ for some measurable set $F \subset \mathbb{R}^N$.

By Proposition 5.22, $\phi^{-1}(F)$ is a measurable set and so

$$\phi^{-1}(F) \cap E = \phi^{-1}(F \setminus A) \cap E$$

is measurable. By Theorem 5.11,

$$\int_{\phi^{-1}(F) \cap E} |J_\phi(x)| \, dx = \int_{\mathbb{R}^N} N(\phi, \phi^{-1}(F) \cap E, y) \, dy,$$

i.e.

$$\int_E v \circ \phi(x) |J_\phi(x)| \, dx = \int_{\mathbb{R}^N} N(\phi, E, y) v(y) \, dy. \tag{5.41}$$

It follows that (5.40) holds if $v$ is a linear combination of a finite number of characteristic functions of measurable, bounded sets $E \subset D$.

*Case 2.* Assume that $\bar{E} \subset D$.

Let $\{v_n : n \in \mathbb{N}\}$ be a nondecreasing, nonnegative sequence of simple, mea-surable functions and let $C \subset \mathbb{R}^N$ be a set of measure zero such that

$$\lim_{n \to +\infty} v_n(y) = v(y)$$

for every $y \in \mathbb{R}^N \setminus C$. Since $\bar{E} \subset D$ and $J_\phi \in L^1_{loc}(D)$, by Theorem 5.11 we deduce that

$$N(\phi, E, \cdot) \in L^1(\mathbb{R}^N),$$

so

$$N(\phi, E, y) < +\infty$$

for almost every $y \in \mathbb{R}^N$ and

$$v_n N(\phi, E, \cdot) \to v N(\phi, E, \cdot) \quad \text{in } L^1(\mathbb{R}^N).$$

By Proposition 5.22,
$$\mathcal{L}^N(\phi^{-1}(C) \cap (E \setminus A)) = 0,$$

where $A := \{x \in D : J_\phi(x) = 0\}$, and as $\lim_{n \to +\infty} v_n \circ \phi(x) = v \circ \phi(x)$ for every $x \in E \setminus \phi^{-1}(C)$, we obtain

$$\lim_{n \to +\infty} v_n \circ \phi(x)|J_\phi(x)| = v \circ \phi(x)|J_\phi(x)|$$

for almost every $x \in E \setminus A$. On the other hand, by Case 1 for all $n \in \mathbb{N}$,

$$\int_{E \setminus A} v_n \circ \phi(x)|J_\phi(x)| \, dx = \int_{\mathbb{R}^N} N(\phi, E \setminus A, y) v_n(y) \, dy$$

and since $\{v_n \circ \phi(x)|J_\phi(x)|\}$ and $\{N(\phi, E, y)v_n(y)\}$ are nonnegative and nondecreasing, by the Beppo–Levi Theorem (or the Lebesgue Monotone Convergence Theorem), we deduce that

$$\int_{E \setminus A} v \circ \phi(x)|J_\phi(x)| \, dx = \int_{\mathbb{R}^N} N(\phi, E \setminus A, y), dy,$$

which, together with the fact that $J_\phi = 0$ a.e. in $A$, and using Lemma 5.8, implies

$$\int_E v \circ \phi(x) \mid J_\phi(x) \mid dx = \int_{E \setminus A} v \circ \phi(x) J_\phi(x) \, dx$$

$$= \int_{\mathbb{R}^N} N(\phi, E \setminus A, y) \, dy$$

$$= \int_{\mathbb{R}^N} N(\phi, E, y) \, dy.$$

*Case 3.* General case, $E \subset D$ and $v \in L^\infty(\mathbb{R}^N)$.
There exists a sequence $\{K_k : k \in \mathbb{N}\}$ of compact sets such that

$$K_k \subset K_{k+1} \subset E \quad \text{and} \quad \mathcal{L}^N(E \setminus K_k) \leq \frac{1}{k},$$

for every $k \in \mathbb{N}$. Since

$$\int_{K_k} v \circ \phi(x)|J_\phi(x)|\, dx = \int_{\mathbb{R}^N} N(\phi, K_k, y)\, dy$$

for every $k \in \mathbb{N}$, passing to the limit in $k$ and using the Beppo–Levi Theorem, we deduce that

$$\int_E v \circ \phi(x)|J_\phi(x)|\, dx = \int_{\mathbb{R}^N} N(\phi, E, y)\, dy.$$

$\square$

## 5.4 Change of variables via the degree

**Proposition 5.25** *Let $D \subset \mathbb{R}^N$ be an open, bounded set and let $\phi \in C(D)^N$ be a mapping which has a weak differential almost everywhere and has the N-property. Assume that $J_\phi \in L^1_{loc}(D)$. Then, for every bounded, open set $G \subset\subset D$ satisfying $\mathcal{L}^N(\partial G) = 0$, we have*

(i) $d(\phi, G, \cdot) \in L^1(\mathbb{R}^N)$;

(ii) $\int_G J_\phi(x)\, dx = \int_{\mathbb{R}^N} d(\phi, G, y)\, dy$.

**Proof** Fix $G \subset\subset D$. Given $v \in L^\infty(\mathbb{R}^N)$, set

$$N_\phi(v, G, y) = \begin{cases} \displaystyle\sum_{x \in \phi^{-1}(y) \cap G} v(x) & \text{if } N(\phi, G, y) < +\infty \\ +\infty & \text{if } N(\phi, G, y) = +\infty. \end{cases}$$

*Claim 1.* $N_\phi(v, D, y) = N_\phi(w, D, y)$ $\mathcal{L}^N$ a.e. $y \in \mathbb{R}^N$ and for every $w \in L^\infty(D)$ such that $v(x) = w(x)$ $\mathcal{L}^N$ a.e. $x \in \mathbb{R}^N$.

Indeed, let

$$A := \{y \in \mathbb{R}^N : N(\phi, G, y) = +\infty\}.$$

By Theorem 5.11 we obtain

$$\mathcal{L}^N(A) = 0.$$

Let $B \subset D$ be such that

$$\mathcal{L}^N(B) = 0 \text{ and } v(x) = w(x)$$

for every $x \in D \setminus B$. Since $\phi$ has the N-property, $\mathcal{L}^N(\phi(B)) = 0$, and for every $y \in \mathbb{R}^N \setminus (\phi(B) \cup A)$ we have

$$\sum_{x \in \phi^{-1}(y) \cap G} v(x) = \sum_{x \in \phi^{-1}(y) \cap G} w(x),$$

thus

$$N_\phi(v, G, y) = N_\phi(w, G, y)$$

for almost every $y \in \mathbb{R}^N$.

*Claim 2.* $N_\phi(v, G, \cdot) \in L^1(\mathbb{R}^N)$.

Let $\{v_n : n \in \mathbb{N}\}$ be a nondecreasing sequence of simple, measurable functions which converges pointwise to $v$, and let

$$v_n := \sum_{i=1}^{k_n} a_n^i \chi_{E_n^i}.$$

As

$$N_\phi(v_n, G, y) = \sum_{i=1}^{k_n} a_n^i N(\phi, E_n^i, y)$$

by Theorem 5.5, $N_\phi(v_n, G, \cdot)$ is measurable. Furthermore, as $\{N_\phi(v_n, G, \cdot) : n \in \mathbb{N}\}$ is a nondecreasing sequence which converges pointwise to $N_\phi(v, G, \cdot)$, we conclude that $N_\phi(v, G, \cdot)$ is measurable. This, together with the fact that $N(\phi, G, \cdot) \in L^1(\mathbb{R}^N)$ (see Theorem 5.11) and the fact that

$$N_\phi(v, G, \cdot) \leq N(\phi, G, \cdot) \sup_{x \in D} |v(x)|,$$

yields

$$N_\phi(v, G, \cdot) \in L^1(\mathbb{R}^N),$$

proving Claim 2.

Now, by Theorem 5.11, we obtain

$$\int_G v_n(x)|J_\phi(x)|\, dx = \int_{\mathbb{R}^N} N_\phi(v_n, G, y)\, dy$$

and so by the Beppo–Levi Theorem we deduce that

$$\int_G v(x)|J_\phi(x)|\, dx = \int_{\mathbb{R}^N} N_\phi(v, G, y)\, dy. \qquad (5.42)$$

In the sequel, we set

$$v(x) = \mathrm{sgn}\,(J_\phi(x))$$

and (5.42) reduces to

$$\int_G J_\phi(x)\, dx = \int_{\mathbb{R}^N} N_\phi(v, G, y)\, dy. \qquad (5.43)$$

*Claim 3.* We claim that $N_\phi(v, G, y) = d(\phi, G, y)$ for almost every $y \in \mathbb{R}^N$.

Indeed, let

$$E_1 := \{x \in D : \phi \text{ does not have a weak differential at } x\},$$

$$E_2 := \{x \in D \setminus E_1 : J_\phi(x) = 0\},$$

and

$$C := A \cup \phi(E_1) \cup \phi(E_2) \cup \phi(\partial G).$$

Since $\mathcal{L}^N(E_1) = \mathcal{L}^N(\partial G) = 0$ and $\phi$ has the $N$-property, we deduce that

$$\mathcal{L}^N(\phi(E_1)) = \mathcal{L}^N(\phi(\partial G)) = 0.$$

Since $\bar{G} \subset D$ is a compact set, by Theorem 5.6 we obtain

$$\mathcal{L}^N(\phi(E_2)) = 0$$

and so

$$\mathcal{L}^N(C) = 0.$$

Let $y \in \mathbb{R}^N \setminus C$. If $\phi^{-1}(y) \cap G = \emptyset$, then $\phi^{-1}(y) \cap \bar{G} = \emptyset$ and by the excision property of the degree (see Theorem 2.7), we have

$$d(\phi, G, y) = 0 = N_\phi(v, G, y).$$

If $\phi^{-1}(y) \cap G = \{a_1, \ldots, a_k\}$, by Lemma 5.10, we deduce that there exists $r > 0$ such that $B(a_1, r), \ldots, B(a_k, r) \subset\subset G$ are mutually disjoint and

$$d(\phi, B(a_i, r), y) = \mathrm{sgn}(J_\phi(a_i)) \quad (i = 1, \ldots, k). \tag{5.44}$$

By the decomposition and the excision property of the degree (see Theorem 2.7) we obtain

$$d(\phi, G, y) = \sum_{i=1}^k d(\phi, B(a_i, r), y) = N_\phi(v, G, y)$$

and so

$$d(\phi, G, y) = N_\phi(v, G, y) \tag{5.45}$$

for almost every $y \in \mathbb{R}^N$. This, together with Claim 2, yields

$$d(\phi, G, \cdot) \in L^1(\mathbb{R}^N)$$

and using (5.43) and (5.45) we conclude that

$$\int_G J_\phi(x)\, dx = \int_{\mathbb{R}^N} d(\phi, G, y)\, dy.$$

$\square$

**Remark 5.26** Let $D$, $\phi$, and $G$ be as in Proposition 5.25.

(i) (5.44) yields

$$d(\phi, G, y) = \sum_{x \in \phi^{-1}} \text{sgn}(J_\phi(x))$$

for almost every $y \in \mathbb{R}^N$ and so

$$|d(\phi, G, y)| \leq N(\phi, G, y)$$

for almost every $y \in \mathbb{R}^N$.

(ii) We observe that if we assume in the proof that $\mathcal{L}^N(\phi(\partial G)) = 0$, without assuming that $\mathcal{L}^N(\partial G) = 0$, the conclusion of Proposition 5.25 still holds.

Next, we state and prove a change of variables formula using the degree function.

**Theorem 5.27** *Let $D \subset \mathbb{R}^N$ be an open set, let $\phi \in C(D)^N$ be such that $\phi$ has the $N$-property, $\phi$ has a weak differential almost everywhere, and $J_\phi \in L^1_{loc}(D)$. Assume that $G \subset D$ is a bounded, open set and that $v \in L^\infty(\mathbb{R}^N)$. Then*

(i) *$y \mapsto v(y)d(\phi, G, y)$ is integrable,*
(ii) *$v \circ \phi\, J_\phi$ is locally integrable,*
(iii) *$\int_G v \circ \phi(x)J_\phi(x)\, dx = \int_{\mathbb{R}^N} v(y)d(\phi, G, y)\, dy$.*

**Proof** By Remark 5.26, $|d(\phi, G, y)| \leq N(\phi, G, y)$ $\mathcal{L}^N$ a.e. $y \in \mathbb{R}^N$ and, together with Theorem 5.23, we conclude (ii). By Proposition 5.25, (i) and (iii) hold in the case where $v$ is a characteristic function. Let $v \in L^\infty(\mathbb{R}^N)$ and we may assume without loss of generality that $v \geq 0$ $\mathcal{L}^N$ a.e. Let $\{v_n : n \in \mathbb{N}\}$ be a nondecreasing sequence of simple functions, bounded in $L^\infty(\mathbb{R}^N)$, and converging pointwise to $v$. Let $A \subset \mathbb{R}^N$ be such that $\mathcal{L}^N(A) = 0$ and

$$\lim_{n \to +\infty} v_n(y) = v(y)$$

for every $y \in \mathbb{R}^N \setminus A$. Setting

$$B := \{x \in D : J_\phi(x) = 0\},$$

we observe that

$$\phi^{-1}(A) = (\phi^{-1}(A) \cap B) \cup (\phi^{-1}(A) \setminus B)$$

and by Theorem 5.11 we obtain

$$\mathcal{L}^N(\phi^{-1}(A) \setminus B) = 0.$$

Clearly,

$$\lim_{n \to +\infty} v_n \circ \phi(x)J_\phi(x) = v \circ \phi(x)J_\phi(x)$$

for every $x \in (\phi^{-1}(A) \cap B) \cup (D \setminus \phi^{-1}(A))$, i.e. $\{v_n \circ \phi J_\phi\}$ converges to $v \circ \phi J_\phi$ almost everywhere. Since $\{v_n\}$ is a sequence bounded in $L^\infty(\mathbb{R}^N)$ and $J_\phi \in L^1_{loc}(D)$, we have

$$v_n \circ \phi(x)J_\phi \to v \circ \phi(x)J_\phi \quad \text{in } L^1(G) \tag{5.46}$$

and, by Proposition 5.25,

$$v_n \, d(\phi, G, \cdot) \to v \, d(\phi, G, \cdot) \quad \text{in } L^1(\mathbb{R}^N). \tag{5.47}$$

Suppose, first, that $v = \chi_V$, where $V \subset D$ is an open set such that $\mathcal{L}^N(\partial V) = 0$. Set

$$F := G \cap \phi^{-1}(V) \cap G.$$

Then $F$ is an open set, $F \subset\subset D$, and

$$\mathcal{L}^N(\phi(\partial F)) \leq \mathcal{L}^N(\phi(\partial G)) + \mathcal{L}^N(\partial V) = 0$$

and so, by Remark 5.26 (i) and (ii) we obtain

$$\int_G v \circ \phi(x)J_\phi(x)\, dx = \int_F J_\phi(x)\, dx$$

$$= \int_{\mathbb{R}^N} d(\phi, F, y)\, dy$$

$$= \int_{\mathbb{R}^N} v(y)d(\phi, F, y)\, dy$$

$$= \int_{\mathbb{R}^N} v(y)d(\phi, G, y)\, dy.$$

Next, assume that $v = \chi_V$ where $V \subset D$ is an arbitrary measurable set. Let $V_\epsilon$ be a family of open sets such that $V \subset V_\epsilon$, $\mathcal{L}^N(V_\epsilon \setminus V) \to 0$ as $\epsilon \to 0^+$. By Lebesgue's Dominated Convergence Theorem and by the previous case,

$$\int_G \chi_V \circ \phi(x)J_\phi(x)\, dx = \lim_{\epsilon \to 0^+} \int_G \chi_{V_\epsilon} \circ \phi(x)J_\phi(x)\, dx$$

$$= \lim_{\epsilon \to 0^+} \int_{\mathbb{R}^N} \chi_{V_\epsilon}(y)d(\phi, G, y)\, dy$$

$$= \int_{\mathbb{R}^N} \chi_V(y)d(\phi, G, y)\, dy$$

since $\chi_{V_\epsilon}(y) \to \chi_V(y)$ for $\mathcal{L}^N$ a.e. $y \in \mathbb{R}^N$ and $d(\phi, G, \cdot) \in L^1(\mathbb{R}^N)$ (see Proposition 5.25).

Finally, for a general $v \in L^\infty(\mathbb{R}^N)$, by (5.46), (5.47) and the latter case,

$$\int_G v \circ \phi(x)J_\phi(x)\, dx = \lim_{n \to +\infty} \int_G v_n \circ \phi(x)J_\phi(x)\, dx$$

$$= \lim_{n \to +\infty} \int_{\mathbb{R}^N} v_n(y) d(\phi, G, y) \, dy$$

$$= \int_{\mathbb{R}^N} v(y) d(\phi, G, y) \, dy.$$

$\square$

## 5.5   Change of variables for Sobolev functions

Here we apply the theory of the last two sections to Sobolev functions. The first theorem is due to Marcus and Mizel (1973).

**Theorem 5.28** *Let $M \geq N$ be two positive integers, $p > N$, let $D \subset \mathbb{R}^N$ be an open bounded set, and let $\phi \in W^{1,p}(D)^M$. Assume that $\phi$ coincides with its continuous representative. Then, for every $A \subset \mathbb{R}^N$ measurable set, we have*

$$H^N(\phi(A)) \leq C(N,p)\mathcal{L}^N(A)^\alpha \left[ \int_A |\nabla \phi(x)|^p \, d\mathcal{L}^N(x) \right]^{\frac{N}{p}},$$

*where $\alpha = 1 - \frac{N}{p}$ and $C(N,p)$ is a constant depending only on $N$ and $p$.*

**Proof**  We may assume, without loss of generality, that $A$ is an open set and let $\{Q_j : j \in \mathbb{N}\}$ be a partition of $A$ into half-open cubes. Using the Sobolev Imbedding Theorem for $W^{1,p}$ (see Theorem 4.45), we obtain for each of these cubes the estimate

$$\max_Q |\phi_i(x) - \bar{\phi}_i| \leq C'(N,p) r^{1-\frac{N}{p}} \left[ \int_Q |\nabla \phi_i(x)|^p \, d\mathcal{L}^N(x) \right]^{\frac{1}{p}},$$

where $\bar{\phi}_i$ is the mean value of $\phi_i$ over $Q$, $r$ is the edge length of $Q$, and $C'(N,p)$ is a constant depending only on $N$ and $p$. Therefore, $\phi(Q_j)$ is contained in the cube centred at $(\bar{\phi}_1, \ldots, \bar{\phi}_M)$ and having edges of length

$$l_j = 2C'(N,p) r_j^{1-\frac{N}{p}} \left[ \int_Q |\nabla \phi_i(x)|^p \, d\mathcal{L}^N(x) \right]^{\frac{1}{p}}$$

and so $\phi(Q_j)$ is contained in the ball $B_j$ of diameter

$$l_j = 2\sqrt{N} C'(N,p) r_j^{1-\frac{N}{p}} \left[ \int_Q |\nabla \phi_i(x)|^p \, d\mathcal{L}^N(x) \right]^{\frac{1}{p}}.$$

Since $l_j$ tends to 0 when $r_j$ tends to zero, we obtain

$$H^N(\phi(A)) \leq C(N,p) \sum_{j=1}^{\infty} r_j^{N(1-\frac{N}{p})} \left[ \int_{Q_j} |\nabla \phi(x)|^p \, d\mathcal{L}^N(x) \right]^{\frac{N}{p}}$$

$$\leq C(N,p) \left[ \sum_{j=1}^{\infty} r_j^N \right]^{1-\frac{N}{p}} \left[ \sum_{j=1}^{\infty} \int_{Q_j} |\nabla \phi(x)|^p \, d\mathcal{L}^N(x) \right]^{\frac{N}{p}}$$

$$= C(N,p)\mathcal{L}^N(A)^{1-\frac{N}{p}} \left[ \int_A |\nabla \phi(x)|^p \, d\mathcal{L}^N(x) \right]^{\frac{N}{p}}.$$

$\square$

**Remark 5.29** If $\phi \in W^{1,p}(D)^N$, $p > N$, then by the Sobolev Imbedding Theorem (Theorem 4.45) and by Theorems 5.21 and 5.27 we conclude that $\phi$ is continuous, it is weakly differentiable almost everywhere in $D$, and it has the $N$-property.

The following versions of Theorems 5.23 and 5.27 are due to Marcus and Mizel (1973).

**Theorem 5.30** Let $D \subset \mathbb{R}^N$ be an open, bounded set, let $p > N$, let $\phi \in W^{1,p}(D)^N$ be equal to its continuous representative, and let $v \in L^{\infty}(\mathbb{R}^N)$. For every measurable set $E \subset D$, we have

$$\int_E v \circ \phi(x)|J_\phi(x)| \, dx = \int_{\mathbb{R}^N} v(y) N(\phi, E, y) \, dy. \tag{5.48}$$

**Proof** By The Sobolev Imbedding Theorem $\phi$ is continuous ($\phi \in C^{0,\alpha}(D)^N$, where $\alpha = 1 - \frac{N}{p}$). By Theorem 5.28, $\phi$ has the $N$-property and by Theorem 5.21, $\phi$ has a weak differential almost everywhere. The result now follows from Theorem 5.23. $\square$

As mentioned in Remark 5.24, Marcus and Mizel (1973) proved Theorem 5.30 for $\phi \in W^{1,p}(D)^m$, $m \geq N$.

**Theorem 5.31** Let $D \subset \mathbb{R}^N$ be an open, bounded set, let $p > N$, and let $\phi \in W^{1,p}(D)^N$ be equal to its continuous representative. If $v \in L^{\infty}(\mathbb{R}^N)$, then, for every open set $G \subset D$ such that $\mathcal{L}^N(\partial G) = 0$, we have

$$\int_G v \circ \phi(x) J_\phi(x) \, dx = \int_{\mathbb{R}^N} v(y) d(\phi, G, y) \, dy.$$

**Proof** The result follows from Remark 5.29 and Theorem 5.27. $\square$

We will focus now on functions in $W^{1,N}$. Due to Corollary 5.19, in what follows we identify $\phi \in W^{1,N}(D)^N$, $J_\phi(x) > 0$ a.e. $x \in D$, with its continuous representative.

**Theorem 5.32** Let $D \subset \mathbb{R}^N$ be an open, bounded set and let $\phi \in W^{1,N}(D)^N$ be a mapping such that $J_\phi(x) > 0$ $\mathcal{L}^N$ a.e. $x \in D$. Then $\phi$ has the $N$- and the $N^{-1}$-properties.

**Proof**

(i) We start by showing that $\phi$ has the $N$-property. Let $a \in D$, $R_a > 0$, be such that

$$Q_{R_a} = (a_1 - R_a, a_1 + R_a) \times \ldots \times (a_N - R_a, a_N + R_a) \subset\subset D.$$

Using the argument of the proof of Theorem 5.16, we observe that for each $i \in \{1, \ldots, N\}$ there exists $I_i \subset (-R_a, R_a)$ such that $\mathcal{L}^1(I_i) = 0$ and

$$\phi|_{P_r^i(a)} \in W^{1,N}(P_r^i(a))^N \cap C(\bar{P}_r^i(a))^N$$

for every $r \in (-R_a, R_a) \setminus I_i$, where

$$
\begin{aligned}
P_r^i(a) = \; & (a_1 - R_a, a_1 + R_a) \times \ldots \times (a_{i-1} - R_a, a_{i-1} + R_a) \times \{a_i + r\} \\
& \times (a_{i+1} - R_a, a_{i+1} + R_a,) \times \ldots \times (a_N - R_a, a_N + R_a).
\end{aligned}
$$

By Theorem 5.28 we obtain

$$H^{N-1}(\phi(P_r^i(a))) < +\infty$$

for every $r \in (-R_a, R_a) \setminus I_i$ and so, by Lemma 4.28, and Remark 4.26, we deduce that

$$\mathcal{L}^N(\phi(P_r^i(a))) = 0 \tag{5.49}$$

for every $r \in (-R_a, R_a) \setminus I_i$.

Let $\theta : \mathbb{R}^N \to \mathbb{R}$ be a symmetric mollifier (see Definition 1.16), such that $0 \le \theta(x)$ for every $x \in \mathbb{R}^N$, $\mathrm{spt}(\theta) \subset B(0,1)$, and for $\epsilon > 0$ define

$$\theta_\epsilon(x) := \frac{1}{r^N} \theta\left(\frac{x}{\epsilon}\right), x \in \mathbb{R}^N, \epsilon > 0, \quad \text{and} \quad \phi_n := \theta_{\frac{1}{n}} * \phi.$$

Then $\phi_n \in C^\infty(D)$,

$$\phi_n \to \phi \quad \text{in } W^{1,N}(D)^N$$

and, by Corollary 5.19,

$$\phi_n \to \phi \quad \text{uniformly in } \bar{Q}_{R_a}.$$

By (5.49) there exists $I \subset (0, R_a)$ such that $\mathcal{L}^1(I) = 0$ and

$$\mathcal{L}^N(\phi(\partial Q_r(a))) = 0 \tag{5.50}$$

for every $r \in (0, R_a) \setminus I$. Fix $r \in (0, R_a) \setminus I$.

*Claim 1.* For every $b \in \phi(Q_r(a)) \setminus \phi(\partial Q_r(a))$ we have $d(\phi, Q_r(a), b) \ge 1$.
Let $b \in \mathbb{R}^N \setminus \phi(\partial Q_r(a))$ and $0 < \epsilon < \mathrm{dist}(b, \phi(\partial Q_r(a)))$. Since $\phi_n$ converges uniformly to $\phi$ in $\bar{Q}_{R_a}$, for $n$ large enough, and by Proposition 1.7 we have

$$d(\phi_n, Q_r(a), b) = \int_{\mathbb{R}^N} \theta_\epsilon(\phi_n(x) - b) J_{\phi_n}(x)\, dx.$$

Letting $n \to +\infty$ and by Theorem 2.3 we deduce that

$$d(\phi, Q_r(a), b) = \int_{\mathbb{R}^N} \theta_\epsilon(\phi(x) - b) J_\phi(x) \, dx,$$

and since $J_\phi(x) > 0$ $\mathcal{L}^N$ a.e. $x \in D$ and $d(\phi, Q_r(a), b)$ is an integer number, by (5.50) we deduce that

$$d(\phi, Q_r(a), b) \geq 1$$

for $\mathcal{L}^N$ almost every $b \in \mathbb{R}^N$.

*Claim 2.* $\chi_n$ converges to $\chi$ $\mathcal{L}^N$ almost everywhere, where $\chi_n$ is the characteristic function of $\phi_n(Q_r(a))$ and $\chi$ is the characteristic function of $\phi(Q_r(a))$.

Indeed, let $b \in \mathbb{R}^N \setminus \phi(\partial Q_r(a))$. If $b \notin \phi(\bar{Q}_r(a))$ there exists $n(x, \epsilon) \in \mathbb{N}$ such that

$$|\phi_n(x) - \phi(x)|_2 < \text{dist}(b, \phi(\bar{Q}_r(a)) \quad \text{for all } x \in \bar{Q}_r(a)$$

for every $n \geq n(x, \epsilon)$. Therefore, $b \notin \phi_n(\bar{Q}_r(a))$ for every $n \geq n(x, \epsilon)$ and so

$$\lim_{n \to +\infty} \chi_n(b) = \chi(b).$$

If $b \in \phi(\bar{Q}_r(a)) \setminus \phi(\partial Q_r(a))$, by the result of Claim 1 we have $d(\phi, Q_r(a), b) \geq 1$ and by Theorem 2.1, $\chi(b) = 1$. Using the fact that $d(\phi, Q_r(a), b) = d(\phi_n, Q_r(a), b)$ for $n$ large enough, and again by Theorem 2.1, we deduce that $\chi_n(b) = 1$ and so

$$\lim_{n \to +\infty} \chi_n(b) = \chi(b).$$

Recalling that $\mathcal{L}^N(\phi(\partial Q_r(a))) = 0$ (see (5.50)), we obtain that $\chi_n$ converges to $\chi$ $\mathcal{L}^N$ almost everywhere.

*Claim 3.* $\mathcal{L}^N(\phi(Q_r(a))) \leq \int_{Q_r(a)} J_\phi(x) \, dx$ for every $r \in (0, R_a)$.

If $r \in (0, R_a) \setminus I$, using Fatou's Lemma, Theorem 5.31, and the result of Claim 2, we obtain

$$\mathcal{L}^N(\phi(Q_r(a))) = \int_{\mathbb{R}^N} \chi(x) \, dx$$

$$\leq \liminf_{n \to +\infty} \int_{\mathbb{R}^N} \chi_n(x) \, dx$$

$$\leq \liminf_{n \to +\infty} \int_{Q_r(a)} J_{\phi_n}(x) \, dx$$

$$= \int_{Q_r(a)} J_\phi(x) \, dx. \tag{5.51}$$

The result for arbitrary $r \in (0, R_a)$ follows from (5.51) and the Lebesgue Monotone Convergence Theorem, where it suffices to consider an increasing sequence $r_k \in (0, R_a) \setminus I$ such that $r_k \to r$.

*Claim 4.* $\phi$ has the $N$-property.

Let $A \subset D$ be such that $\mathcal{L}^N(A) = 0$. We may assume without loss of generality that $A \subset\subset D$. Fix $\epsilon > 0$. Since $J_\phi \in L^1(D)$ there exists an open set $O \subset\subset D$ cointaining $A$ such that

$$\int_O J_\phi(x)\, dx < \epsilon$$

and there exists a collection $\{Q_j : j \in \mathbb{N}\}$ of half-open cubes such that $\{Q_j : j \in \mathbb{N}\}$ is a partition of $O$ and

$$O = \bigcup_{i \in \mathbb{N}} Q_j$$

(see Rudin 1966, Section 2.19). By (5.51) we have

$$\mathcal{L}^N(\phi(A)) \leq \mathcal{L}^N(\phi(O))$$
$$\leq \sum_{i=1}^\infty \mathcal{L}^N(\phi(Q_j))$$
$$\leq \sum_{i=1}^\infty \int_{Q_j} J_\phi(x)\, dx$$
$$= \int_O J_\phi(x)\, dx \leq \epsilon.$$

It suffices to let $\epsilon \to 0^+$.

(ii) We show now that $\phi$ has the $N^{-1}$-property. In fact, by part (i) and by Theorems 5.21 and 5.11, we have

$$\int_E J_\phi(x)\, dx = \int_{\mathbb{R}^N} N(\phi, E, y)\, dy \qquad (5.52)$$

for every $E \subset D$. Thus, setting $E = \phi^{-1}(F)$ where $\mathcal{L}^N(F) = 0$, we obtain

$$\int_{\phi^{-1}(F)} J_\phi(x)\, dx = \int_{\mathbb{R}^N} N(\phi, \phi^{-1}(F), y)\, dy$$
$$= \int_F N(\phi, \phi^{-1}(F), y)\, dy$$
$$= 0$$

and as $J_\phi(x) > 0$ for $\mathcal{L}^N$ a.e. $x \in D$, we conclude that $\mathcal{L}^N(\phi^{-1}(F)) = 0$.

$\square$

**Remark 5.33** It can be shown that if $\phi \in W^{1,N}(D)^N$ is continuous, open and *discrete* (i.e. $\phi^{-1}\{y\}$ is finite, for all $y \in \mathbb{R}^N$), then $\phi$ satisfies the $N$-property. For details we refer the reader to Martio and Ziemer (1992).

The two following theorems were proved by Gold'sthein and Reshetnyak (1990), where, as usual, due to Corollary 5.19 we assume that $\phi$ coincides with its continuous representative.

**Theorem 5.34** *Let $D \subset \mathbb{R}^N$ be an open, bounded set and let $\phi \in W^{1,N}(D)^N$ be a mapping such that $J_\phi > 0 \; \mathcal{L}^N$ a.e. $x \in D$. If $v \in L^\infty(\mathbb{R}^N)$ then, for every measurable set $E \subset D$, we have*

$$\int_E v \circ \phi(x) |J_\phi(x)| \, dx = \int_{\mathbb{R}^N} v(y) N(\phi, E, y) \, dy.$$

**Proof** By Theorem 5.21 $\phi$ has a weak differential almost everywhere and, by Theorem 5.32, $\phi$ has the $N$-property. The result now follows from Theorem 5.23.
□

**Theorem 5.35** *Let $D \subset \mathbb{R}^N$ be an open, bounded set and let $\phi \in W^{1,N}(D)^N$ be a mapping such that $J_\phi > 0 \; \mathcal{L}^N$ a.e. $x \in D$. If $v \in L^\infty(\mathbb{R}^N)$, then, for every open set $G \subset D$ such that $\mathcal{L}^N(\partial G) = 0$, we have*

$$\int_G v \circ \phi(x) J_\phi(x) \, dx = \int_{\mathbb{R}^N} v(y) d(\phi, G, y) \, dy. \tag{5.53}$$

**Proof** By Corollary 5.19 and Theorem 5.21, $\phi$ is continuous and has a weak differential almost everywhere and, by Theorem 5.32, $\phi$ has the $N$-property. It suffices to apply Theorem 5.27.
□

Recently, Švérak (1988) and Müller *et al.* (1994), generalized Theorems 5.34 and 5.35 to a class of Sobolev mappings not necessarily continuous but under some additional regularity hypotheses on the boundary. We recall that the notion of Lipschitz boundary was introduced in Definition 4.40. Although the space $\mathcal{A}^+_{p,q}(D)$ has already been defined in Section 5.2, for convenience we include its description below.

**Definition 5.36** *Let $D \subset \mathbb{R}^N$ be an open set with Lipschitz boundary, let $p \geq N - 1$, $q \geq \frac{N}{N-1}$. Then*

$$\mathcal{A}_{p,q}(D) := \{\phi \in W^{1,p}(D)^N \; : \; \mathrm{adj}(\nabla\phi) \in L^q(D)^{N \times N}\}$$

*and*

$$\mathcal{A}^+_{p,q}(D) = \{\phi \in \mathcal{A}_{p,q}(D) \; : \; J_\phi(x) > 0 \; \mathcal{L}^N \text{ a.e. } x \in D\}.$$

We recall that if $p > N - 1$ mappings in $\mathcal{A}^+_{p,q}(D)$ are continuous outside sets of Hausdorff dimension less than or equal to $N - p$ (see Remark 5.18). Also, if $\phi \in \mathcal{A}_{p,q}(D)$, then $J_\phi \in L^1(D)$. In fact, from the identity

$$J_\phi(x)I_N = \nabla\phi(x)\,\mathrm{adj}\,(\phi(x))$$

we obtain

$$|J_\phi(x)|^N = |J_\phi(x)|\,|\det\mathrm{adj}(\phi(x))| \le |J_\phi(x)|\,|\mathrm{adj}(\phi(x))|^N$$

and so

$$|J_\phi(x)| \le |\,\mathrm{adj}\,(\phi(x))|^{\frac{N}{N-1}} \in L^1(D).$$

We use $\wedge$ to denote the exterior product (or wedge product) between vectors and we recall that the mapping

$$(\xi_1,\dots,\xi_{N-1}) \in \left(\mathbb{R}^N\right)^{N-1} \mapsto \xi_1 \wedge \dots \wedge \xi_{N-1}$$

is multilinear, alternate and, if $\{e_1,\dots,e_N\}$ is the canonical basis of $\mathbb{R}^N$, then

$$e_1 \wedge \dots \wedge e_{i-1} \wedge e_{i+1} \wedge \dots \wedge e_N = (-1)^{N-1}e_i, \quad i = 1,\dots,N.$$

It can easily be verified that if $F$ is a $N \times N$ matrix, then

$$|\mathrm{adj}\,F| = \sup\left\{|(F\xi_i) \wedge \dots \wedge (F\xi_{N-1})| : |\xi_i \wedge \dots \wedge \xi_{N-1}| \le 1\right\}. \tag{5.54}$$

**Definition 5.37** *A function $\varphi : \partial D \to \mathbb{R}$ belongs to $L^p(\partial D)$ (resp. $W^{1,p}(\partial D)$) if $\varphi_i : y_i' \mapsto \varphi(y_i', a_i(y_i'))$ belongs to $L^p(\partial D)$ (resp. $W^{1,p}(\partial D)$), for all $i = 1,\dots,M$.*

If $\varphi \in W^{1,p}(\partial D)$, then $d\varphi$ is a 1-form given, in local coordinates, by

$$d\varphi = \sum_{j=1}^{N-1} \frac{\partial\varphi_i}{\partial y_i'^j} dy_i'^j, \quad i = 1,\dots,M.$$

Suppose now that $\varphi \in W^{1,p}\left(\partial D; \mathbb{R}^N\right)$, let $x \in \partial D$, and let $T_x$ denote the tangent space to $\partial D$ at $x$. If $\xi \in T_x$ and if $\{e_1,\dots,e_N\}$ is the canonical basis of $\mathbb{R}^N$, then

$$d\varphi(x)(\xi) := \sum_{k=1}^{N} d\varphi^k(x)(\xi)e_k.$$

We define

$$|\mathrm{adj}_s\varphi|(x) := \sup\{|d\varphi(x)(\xi_i) \wedge \dots \wedge d\varphi(x)(\xi_{N-1})| :$$
$$\xi_1,\dots,\xi_{N-1} \in T_x, |\xi_1 \wedge \dots \wedge \xi_{N-1}| \le 1\}$$

and

$$\mathcal{A}_{p,q}(\partial D) := \left\{\varphi \in W^{1,p}(\partial D) : \int_{\partial D} |\mathrm{adj}_s\varphi|^q(x)\,dH^{N-1}(x) < +\infty\right\}.$$

Next we show that we can define $d(\phi,\partial D,\cdot)$ in a natural way for functions $\varphi \in W^{1,p}(\partial D)^N \cap C(\partial D)^N, \quad p \ge N - 1$.

Let $\Psi \in C^\infty(\bar{D})^N$, let $y_0 \in \mathbb{R}^N \setminus \Psi(\partial D)$, and let $f \in C^\infty(\mathbb{R}^N)^N$ be such that spt $f$ is contained in the connected component of $\mathbb{R}^N \setminus \Psi(\partial D)$ which contains $y_0$. Choose $g \in C^\infty(\mathbb{R}^N)^N$ such that div $g = f$ (e.g. $g_i := C\frac{x_i}{|x|^N} * f$, for a suitable constant $C$) and recall that

$$\sum_{j=1}^N \frac{\partial}{\partial x_j}(\mathrm{adj}\nabla\Psi)_{ij} = 0, \quad i = 1, \dots, N$$

(see Exercise 1.3). By Theorems 2.3, 5.27, and Remark 1.14, we have

$$d(\Psi, D, y_0) \int_{\mathbb{R}^N} f(y)\, dy = \int_D f(\Psi(x)) J_\Psi(x)\, dx$$

$$= \int_D \sum_{i,j=1}^N \frac{\partial}{\partial x_j}[g_i(\Psi(x))(\mathrm{adj}\nabla\Psi(x))_{ij}]\, dx$$

$$= \int_{\partial D} \sum_{i,j=1}^N g_i(\Psi(x))(\mathrm{adj}\nabla\Psi(x))_{ij}\nu_j(x)\, dH^{N-1}(x),$$

where $\nu$ denotes the outward unit normal to $\partial D$ at $x$ (which exists $H^{N-1}$ a.e. because $\partial D$ is Lipschitz). Using the notation introduced in Chapter 1, Section 3, we can write

$$d(\Psi, D, y_0) \int_{\mathbb{R}^N} f(y)\, dy = \int_D \Psi^*(f\, dy_1 \wedge \dots \wedge dy_N)$$

$$= \int_D \Psi^*(d\beta) = \int_D d\Psi^*(\beta) = \int_{\partial D} \Psi^*(\beta), \quad (5.55)$$

where we have used Stokes' Theorem, and where

$$\beta := \sum_{i=1}^N (-1)^{i-1} g_i dy_1 \wedge \dots \wedge \hat{dy}_{i-1} \wedge dy_{i+1} \wedge \dots \wedge dy_N.$$

Now let $\phi \in W^{1,p}(\partial D)^N \cap C(\partial D)^N$, $p \geq N - 1$, and $y_0 \in \mathbb{R}^N \setminus \Psi(\partial D)$. By Theorem 1.12 and (5.55), $d(\phi, \partial D, y_0)$ will be well defined and (5.55) will hold for $\phi$ if we show that we may find a sequence $\Psi_n \in C^\infty(\bar{D})^N$ such that $\Psi_n|_{\partial D} \to \phi|_{\partial D}$ uniformly and $\Psi_n \to \phi$ in $W^{1,p}(\partial D)^N$. To construct the approximating sequence $\{\Psi_n : n \in \mathbb{N}\}$, using a partition of unity and a change of coordinates we may assume that there exist $\alpha, \beta > 0$ and a Lipschitz function $a$ such that

$$a(0) = 0, \quad \|a\| \leq \frac{\beta}{4}, \quad D \cap \mathcal{U} = \{(x', x_N) : a(x') < x_N < a(x') + \beta, x' \in U\}$$

and

$$U = (-\alpha, \alpha)^{N-1}, \ \mathcal{U} = \{(x', x_N) : x' \in U, \ |a(x') - x_N| < \beta\}.$$

Let

$$v(x') := \phi(x', a(x')).$$

Then $v \in W^{1,p}(U)^N$ and, by the Sobolev Imbedding Theorem, and as spt $v \subset\subset U$, then $v \in C_c(U)^N$ and we may find $v_n \in C_c^\infty(U)^N$ such that $v_n \to v$ uniformly and in $W^{1,p}(U)^N$. Let

$$\Psi_n(x) := \eta(x_N)v_n(x'),$$

where $\eta \in C_c^\infty\left(\frac{-\beta}{2}, \frac{\beta}{2}\right), \eta = 1$ on $\left[\frac{-\beta}{4}, \frac{\beta}{4}\right]$. Then $\Psi_n \in C_c^\infty(\mathbb{R}^N)^N$, and, since $\Psi_n(x', a(x')) = v_n(x')$, we conclude that $\Psi_n|_{\partial D}$ converges to $\phi$ uniformly on $\partial D$ and in $W^{1,p}(D)^N$ strongly.

**Theorem 5.38** *Let $D \subset \mathbb{R}^N$ be an open, bounded set with Lipschitz boundary and let $p \geq N - 1$, $q \geq \frac{N}{N-1}$. Let $\phi \in \mathcal{A}_{p,q}(D)$ be such that its trace $T(\phi) \in \mathcal{A}_{p,q}(\partial D) \cap C(\partial D)^N$, let $y_0 \in \mathbb{R}^N \setminus \phi(\partial D)$, and let $f \in C(\mathbb{R}^N)$ be a bounded function supported in the connected component of $\mathbb{R}^N \setminus T(\phi)(\partial D)$ which contains $y_0$.*

(i)

$$d(T(\phi), \partial D, y_0) \int_{\mathbb{R}^N} f(y)\, dy = \int_D f \circ \phi(x) J_\phi(x)\, dx.$$

*In addition, if $p > N - 1$, then*

(ii)

$$\int_D f \circ \phi(x) J_\phi(x)\, dx = \int_{\mathbb{R}^N} f(y) d(T(\phi), \partial D, y)\, dy.$$

(iii) *If $\phi \in \mathcal{A}_{pq}^+(D)$, then*

$$\int_E f \circ \phi(x) J_\phi(x)\, dx = \int_{\mathbb{R}^N} f(y) N(\phi, E, y)\, dy$$

*for every measurable set $E \subset D$.*

**Remark 5.39**

(i) Theorem 5.38 was first proved by Švérak (1988) for $q \geq q\frac{p}{p-1}$ and later extended by Müller *et al.* (1994) to $q \geq \frac{N}{N-1}$. We refer the reader to these papers for the proof.

(ii) If $p > N - 1$, then $T(\phi)$ has a continuous representative by the Sobolev Imbedding Theorem. In this case, by the Tietze Extension Theorem (see Theorem 1.15) there exists a continuous extension $\tilde{\phi} : \bar{D} \to \mathbb{R}^N$ of $T(\phi)$ and, by Theorem 2.4, given any other continuous extension $\hat{\phi} : \bar{D} \to \mathbb{R}^N$ of $T(\phi)$, we have

$$d(\tilde{\phi}, D, y) = d(\hat{\phi}, D, y)$$

for all $y \in \mathbb{R}^N \setminus T(\phi)(\partial D)$. This defines, in a natural way, $d(T(\phi), \partial D, y)$ for $y \in \mathbb{R}^N \setminus T(\phi)(\partial D)$, in agreement with the result of (5.55).

# 6

# LOCAL INVERTIBILITY OF SOBOLEV FUNCTIONS AND APPLICATIONS

In this chapter we show that $W^{1,N}$ functions with positive jacobian are locally invertible, with inverse in $W^{1,1}$ (see Theorem 6.1). Under suitable growth hypotheses, we are able to improve the regularity of the inverse function to $W^{1,s}$ for some $s > 1$. In addition, we give a necessary condition for Sobolev functions to be invertible.

As an application of the local invertibility property, in Section 6.2 we study the weak lower semicontinuity of functionals $E$ of the form

$$E(u,v) := \int_\Omega W\left(\nabla u(x)(\nabla v(x))^{-1}\right)\, dx,$$

defined on the set

$$B_{p,q} = \left\{(u,v) \in W^{1,p}(\Omega)^N \times W^{1,q}(\Omega)^N : \det \nabla v(x) = 1 \text{ a.e } x \in \Omega\right\},$$

where $1 \le p < +\infty, N \le q \le +\infty$, and $\frac{1}{p} + \frac{N-1}{q} = \frac{1}{r} \le 1$. When $N = 3$, $\nabla u(x) \cdot (\nabla v(x))^{-1}$ represents the lattice of a neutral elasto-plastic change of state of a perfect cubic crystal, $u$ is the elastic deformation, and $v$ corresponds to the slip or plastic deformation (for details, see Ericksen 1987, Davini and Parry 1989, Fonseca and Parry 1987, and Dacorogna and Fonseca 1992). We prove that, under some convexity and growth assumptions on the function $W$, $E$ is weakly lower semicontinuous on $B_{p,q}$. If $r > 1$ and $q \ne +\infty$ we rely on the div–curl lemma (see Tartar 1979) to guarantee that

$$\nabla u_\epsilon \cdot (\nabla v_\epsilon)^{-1} \rightharpoonup \nabla u \cdot (\nabla v)^{-1} \text{weakly in } L^r$$

whenever

$$(u_\epsilon, v_\epsilon) \rightharpoonup (u,v) \quad quadweakly \text{ in } B_{p,q}.$$

## 6.1   Local invertibility in $W^{1,N}$

The result given in this section is in the same spirit as the work of Ball (1981), Ciarlet and Nečas (1987), Švérak (1988), and Tang (1988). As far as we know, the existence and regularity of the local inverse function $w$ is not an immediate consequence of these earlier results, where assumptions are placed either on the trace $v|_{\partial\Omega}$ or on $\mathcal{L}^N(v(\Omega))$. Using Lemma 5.9 and the invertibility result provided by Tang (1988), we obtain the existence of the local inverse function $w$ and

149

then deduce its regularity. Due to his relaxed assumption $q > N - 1$ (here we have $q \geq N$), Tang used an elaborate method to obtain the existence of an inverse $w \in W^{1,1}_{loc}$ under the condition introduced by Ciarlet and Nečas (1987), $\int_\Omega \det \nabla v(x)\, dx \leq \mathcal{L}^N(v(\Omega))$.

The proof presented here concerning the local invertibility of $v$ is independent of the work by Ball (1981), Ciarlet and Nečas (1987), Šverák (1988), and Tang (1988), and the method employed relies on the basic properties of degree theory.

We first state the main result of this section and some of its corollaries. In what follows, due to Corollary 5.19, we identify every function $v \in W^{1,N}(\Omega)^N$ such that $J_v(x) > 0$ $\mathcal{L}^N$ a.e. $x \in D$ with its continuous representative.

**Theorem 6.1** *Let $\Omega \subset \mathbb{R}^N$ be a bounded, open set and let $v \in W^{1,N}(\Omega)^N$ be a function such that $\det \nabla v(x) > 0$ a.e. $x \in \Omega$. Then for almost every $x_0 \in \Omega$, $v$ is locally almost invertible in a neighbourhood of $x_0$, in the sense that there exist $r \equiv r(x_0) > 0$, an open set $D \equiv D(x_0) \subset\subset \Omega$, and a function $w : B(y_0, r) \to D$, with $y_0 = v(x_0)$, such that*

$$w \in W^{1,1}(B(y_0, r))^N,$$
$$w \circ v(x) = x \text{ a.e. } x \in D, \tag{0.1}$$
$$v \circ w(y) = y \text{ a.e. } y \in B(y_0, r), \tag{0.2}$$
$$\nabla w(y) = (\nabla v)^{-1}(w(y)) \text{ a.e. } y \in B(y_0, r). \tag{0.3}$$

*If, in addition, $|\frac{\mathrm{adj}(\nabla v)}{\det \nabla v}|_2^s, \det \nabla v \in L^1(\Omega)$ for some $1 \leq s < +\infty$, then $w \in W^{1,s}(B(y_0, r), D)$.*

Before proving Theorem 6.1 we list some of its consequences, which will be proved at the end of this section.

**Corollary 6.2** *Let $\Omega \subset \mathbb{R}^N$ be a bounded, open set, $q \geq N$, and let $v \in W^{1,q}(\Omega)^N$ be a function such that $\det \nabla v(x) \neq 0$ a.e. $x \in \Omega$.*

(a) *Assume that $\Omega_1, \Omega_2 \subset \mathbb{R}^N$ are two open sets and $N \subset \mathbb{R}^N$ is a set of measure zero such that $\Omega = \Omega_1 \cup \Omega_2 \cup N$, $\det \nabla v(x) > 0$ a.e. $x \in \Omega_1$, and $\det \nabla v(x) < 0$ a.e. $x \in \Omega_2$. Then for almost every $x_0 \in \Omega$, $v$ is locally almost invertible in a neighbourhood of $x_0$ (in the sense of (6.1), (6.2) and (6.3)).*

(b) *Conversely, if $q > N$, $v \in W^{1,q}(\Omega)^N$ and if for almost every $x_0 \in \Omega$ $v$ is locally almost invertible in a neighbourhood of $x_0$, then there are open sets $\Omega_1, \Omega_2 \subset \mathbb{R}^N$ and a null set $E \subset \mathbb{R}^N$ such that $\Omega = \Omega_1 \cup \Omega_2 \cup E$, $\det \nabla v(x) > 0$ a.e. $x \in \Omega_1$, and $\det \nabla v(x) < 0$ a.e. $x \in \Omega_2$.*

**Corollary 6.3** *Let $q \geq N$, let $\Omega \subset \mathbb{R}^N$ be a bounded, open set, and let $v \in W^{1,q}(\Omega)^N$ be a function such that $\det \nabla v(x) = 1$ a.e. $x \in \Omega$. Then the inverse function $w$ (obtained in Theorem 6.1) is such that*

$$w \in W^{1, \frac{q}{N-1}}(v(D))^N.$$

*If, in addition, $q \geq N(N-1)$, then $w \circ v(x) = x$ for every $x \in D$, $v \circ w(y) = y$ for every $y \in B(y_0, r)$, $v$ is a local homeomorphism, and $v$ is an open mapping on $\Omega \setminus L$ for some set $L \subset \Omega$ of measure zero. In particular, if $N = 2$, then $N(N-1) = N = 2$ and $v$ is a local homeomorphism at $x_0$.*

## Remark 6.4

(i) By Theorem 5.32, if $v \in W^{1,N}(\Omega)^N$ with $\det \nabla v(x) > 0$ for $\mathcal{L}^N$ a.e. $x \in \Omega$, then $v$ satisfies the $N$- and the $N^{-1}$-properties. Also, by Theorem 5.32, if $v \in W^{1,p}(\Omega)^N, p > N$, then $v$ satisfies the $N$-property.

(ii) A function $v \in W^{1,N}(\Omega)^N$ is said to be a mapping of *bounded distortion* if $|\nabla v(x)|_2^N \leq K(\det \nabla v(x))$ for $\mathcal{L}^N$ a.e. $x \in \Omega$ and for some constant $K$. It is well known that every mapping of bounded distortion $v \in W^{1,N}(\Omega)^N$ is a homeomorphism locally at almost every point $x_0 \in \Omega$ (see Reshetnyak 1989, Theorem 6.6, p. 187). Moreover, mappings of bounded distortion are open mappings or are constant in $\Omega$ (see Reshetnyak 1989, Theorem 6.4, p. 184). More generally, defining the *dilatation $K$* of $v$ as

$$K(x) := \begin{cases} \frac{|\nabla \phi(x)|_2^N}{\det \nabla \phi(x)} & \det \nabla \phi(x) \neq 0 \\ 1 & \det \nabla \phi(x) = 0, \end{cases}$$

$v$ is said to have *finite dilatation* if $1 \leq K(x) < +\infty$ for $\mathcal{L}^N$ a.e. $x \in \Omega$. Gold'sthein and Vodopyanov (1977) proved that a mapping $v$ of finite dilatation is continuous and its components $v_i$ are weakly monotone. Recently, Heinonen and Koskela (1993) and Manfredi and Villamor (in press) proved that if $v \in W^{1,N}(\Omega)^N$ has finite dilatation, if $v$ is *quasi-light* (i.e. $v^{-1}(\{y\})$ is compact for all $y \in \mathbb{R}^N$), and if $K \in L^{N-1+\epsilon}$ for some $\epsilon > 0$, then $v$ is open and discrete (i.e. $v^{-1}(\{y\})$ is finite for all $y$ in $\mathbb{R}^N$). In particular, by Remark 5.33 $v$ satisfies the $N$-property.

(iii) Note that, even if $v \in C^1(\bar{\Omega})^N$ is such that $\det \nabla v(x) \geq \gamma > 0$ for all $x \in \Omega$, we cannot expect global invertibility of $v$ without any regularity assumptions on the trace of $v$. As an example, consider

$$\Omega := \{(x, y) \in \mathbb{R}^2 : 1 < x^2 + y^2 < 2\}, \quad v(x, y) := (x^2 - y^2, 2xy).$$

For every $(x, y) \in \Omega$ we have $\det \nabla v(x, y) = 4(x^2 + y^2) \geq 4$, although $v(x, y) = v(-x, -y)$ (see also Ball 1981).

(iv) Under the assumptions of Theorem 6.1, we cannot expect $v$ to be locally invertible everywhere. The following example is provided by Ball (1981): let $N \geq 3$ and consider the cylinder

$$\Omega = \{x \in \mathbb{R}^N : 0 \leq R < 1, |x_N| < 2\},$$

where $x = (x_1, \ldots, x_N), R = \sqrt{x_1^2 + \ldots + x_{N-1}^2}$ and $v = (v_1, \ldots, v_N)$ is defined by

$$v_i(x) := R^{-\alpha} x_i, \quad i = 1, \ldots, N-1,$$
$$v_N(x) := R^\beta x_N, \quad |x_N| \le 1,$$
$$v_N(x) := [2(|x_N| - 1) + (2 - |x_N|)R^\beta] \operatorname{sgn} x_N, \quad 1 \le |x_N| \le 2,$$

where $N < q < N(N-1), \frac{1}{N-1} - \frac{1}{q} < \alpha < \frac{N-1}{q}, \beta = \alpha(N-1)$. It was shown by Ball (1981) that $v \in W^{1,q}(\Omega)^N$, $\det \nabla v(x) > 1 - \alpha > 0$ almost everywhere. However, one can easily see that

$$v^{-1}(0) = \{(0, \ldots, 0, \lambda) : \ |\lambda| \le 1\}$$

and so $v$ cannot be locally invertible at any point $(0, \cdots, 0, \lambda)$ for $|\lambda| \le 1$.

(v) Ball (1981) provided an example of a mapping $v \in W^{1,\infty}(\Omega)^2$, $\Omega \subset \mathbb{R}^2$, with $\det \nabla v(x) = 1$ a.e. $x \in \Omega$, for which there is no sequence $v_n \in C^1(\bar{\Omega})^2$ such that $v_n \to v$ uniformly and $J_{v_n}(x) > 0$ a.e. $x \in \Omega$. Therefore, to prove Theorem 6.1 one cannot expect to approximate the function $v$ by a sequence of locally invertible, smooth functions $v_n$.

(vi) Note that for every bounded, open set $\Omega \subset \mathbb{R}^N$, there exist a measurable set $E \subset \Omega$ of nonzero measure and a homeomorphism $v \in W^{1,\infty}(\Omega)^N$ such that $\det \nabla v(x) = 0$ for every $x \in E$. The following example is provided by Martio and Ziemer (1992, Remarks 3.7). Let $E \subset [0,1]$ be a Cantor set of positive one-dimensional measure $0 < 1 - \alpha < 1$ and write

$$v_1(x) = v_1(x_1, \ldots, x_N) = \int_0^{x_1} \chi_{E^c}(t)\, dt$$

where $\chi_{E^c}$ is the characteristic function of the complement of $E$ in $[0,1]$. Setting $v(x) := (v_1(x), x_2, \ldots, x_N)$, it follows that $v \in W^{1,\infty}(\Omega)^N$, $v$ is a homeomorphism of $[0,1]^N$ onto $[0, \alpha] \times [0,1]^{N-1}$, and $\det \nabla v(x) = 0$ a.e. $x \in E \times [0,1]^{N-1}$.

(vii) Due to the previous remark, the assumption $\det \nabla v(x) \ne 0$ a.e. $x \in \Omega$ in Corollary 6.2 is essential.

**Lemma 6.5** Let $\Omega \subset \mathbb{R}^N$ be an open set, let $v \in W^{1,N}(\Omega)^N$ be such that $\det \nabla v(x) > 0$ a.e. $x \in \Omega$. Then for every $x_0 \in \Omega$ such that $v$ is differentiable at $x_0$ (in the classical sense) and $\det \nabla v(x_0) > 0$ there is $R_0 \equiv R_0(x_0)$ such that for every $0 < R < R_0$ the following holds:

$$N(v, B(x_0, R), y) = 1 \ \text{for almost every } y \in C_R, \tag{6.4}$$
$$d(v, B(x_0, R), y) = 1 \ \text{for every } y \in C_R, \tag{6.5}$$
$$d(v, B, y) = 1 \ \text{for every } y \in v(B) \setminus v(\partial B), \tag{6.6}$$

for every nonempty, open set $B \subset v^{-1}(C_R) \cap B(x_0, R)$ such that $\mathcal{L}^N(\partial B) = 0$, where $C_R$ is the connected component of $\mathbb{R}^N \setminus v(\partial B(x_0, R))$ containing $y_0 := v(x_0)$ and $N(v, E, y)$ is the cardinality of the set $\{x \in E : v(x) = y\}$.

**Proof** By Theorem 5.21, $v$ is differentiable at almost every point $x \in \Omega$. Fix $x_0 \in \Omega$ such that $v$ is differentiable at $x_0$ and $\det \nabla v(x_0) > 0$. By Lemma 5.9 there is $R_0 > 0$ such that $B(x_0, R_0) \subset\subset \Omega$ and $d(v, B(x_0, R), y_0) = 1$ for every $0 < R < R_0$ and (6.5) follows from Theorem 2.3. Since $\det \nabla v(x) > 0$ a.e. $x \in \Omega$, Theorems 5.23, 5.27, along with (6.5), yield (6.4). Finally, we prove (6.6). As $v$ satisfies the $N$-property (see Remark 6.4) $\mathcal{L}^N(v(\partial B(x_0, R_0))) = 0$ and $\mathcal{L}^N(v(\partial B)) = 0$. Since $B$ is a nonempty open set, by Theorem 5.32 we have that $\mathcal{L}^N(v(B)) \neq 0$ and so $\mathcal{L}^N(v(B) \setminus v(\partial B)) > 0$. Let $y \in v(B) \setminus v(\partial B)$ and let $C$ be the connected component of $\mathbb{R}^N \setminus v(\partial B)$ containing $y$. As $\mathcal{L}^N(v(\partial B(x_0, R_0))) = 0$ and since $d(v, B, \cdot)$ is a constant on $C$, we may assume without loss of generality that $y \notin v(\partial B(x_0, R_0))$. Let $\rho_\epsilon \in C^\infty(\mathbb{R}^N)$ be such that

$$0 \leq \rho_\epsilon(y), \text{ for all } y \in \mathbb{R}^N, \epsilon > 0$$

$$\frac{1}{2} \leq \rho_\epsilon(y), \text{ for all } y \in B\left(0, \frac{\epsilon}{2}\right)$$

$$\operatorname{spt} \rho_\epsilon \subset B(0, \epsilon), \text{ for all } \epsilon > 0$$

$$\int_{\mathbb{R}^N} \rho_\epsilon(y)\, dy = 1 \text{ for all } \epsilon > 0. \tag{6.7}$$

Since $y \in v(B)$ there exists $x \in B$ such that $y = v(x)$ and by Theorem 2.3 we have

$$\lim_{\epsilon \to 0} \int_B \rho_\epsilon(v(z) - y) \det \nabla v(z)\, dz = d(v, B, y) \tag{6.8}$$

and using the continuity of $v$ at $x$, we deduce that for every $\epsilon > 0$ there exists $\delta > 0$ such that $|v(z) - y|_2 \leq \frac{\epsilon}{2}$ for every $z \in B(x, \delta)$. Recalling that $\det \nabla v(z) > 0$ a.e. $z \in B(x, \delta)$, by (6.7) and (6.8) we obtain

$$d(v, B, y) > 0. \tag{6.9}$$

Finally as the degree $d(v, \cdot, y)$ is a nondecreasing function of the set, using (6.5) and the fact that $B \subset v^{-1}(C_R) \cap B(x_0, R)$ we obtain

$$d(v, B, y) \leq d(v, B(x_0, R), y) = 1 \tag{6.10}$$

which, together with (6.9) and the fact that the degree is an integer number, yields (6.6). $\qquad \square$

**Lemma 6.6** *Let $\Omega, v, R_0, x_0$, be as in Lemma 6.5, (6.4), (6.5). Let $C_{R_0}$ be the connected component of $\mathbb{R}^N \setminus v(\partial B(x_0, R_0))$ containing $y_0 := v(x_0)$. Then for every $r > 0$ such that $B(y_0, r) \subset\subset C_{R_0}$,*

$$v(O) = B(y_0, r), \quad v(\partial O) \subset \partial v(O) = \partial B(y_0, r),$$

*where $O := v^{-1}(B(y_0, r)) \cap B(x_0, R_0) \subset\subset B(x_0, R_0)$.*

**Proof** It is clear that $v(O) \subset B(y_0, r)$. Conversely, if $y \in B(y_0, r)$, by (6.5) $d(v, B(x_0, R_0), y) = 1$ and so by Theorem 2.1, there exists $x \in B(x_0, R_0)$ such that $y = v(x)$, implying $y \in v(O)$.

Let $x \in \partial O$ and let $\{a_n\} \subset O$, $\{b_n\} \subset B(x_0, R_0) \setminus O$ be such that

$$\lim_{n \to +\infty} a_n = \lim_{n \to +\infty} b_n = x.$$

We have $v(a_n) \in v(O)$ and, as $v(O) = v(v^{-1}(B(y_0, r))) = B(y_0, r)$, we observe that $v(b_n) \notin v(O) = B(y_0, r)$. Using the continuity of $v$ at $x$, we have

$$v(x) = \lim_{n \to +\infty} v(a_n) = \lim_{n \to +\infty} v(b_n)$$

and this gives $x \in \partial v(O)$.                                            □

**Lemma 6.7** Let $v \in W^{1,N}(\Omega)^N$, $\det \nabla v(x) > 0$ a.e. $x \in \Omega$ and let $x_0 \in D$ be such that $v(x) \neq v(x_0)$ for every $x \in \bar{B}(x_0, R_0) \setminus \{x_0\}$. For every $0 < R < R_0$ there exists $r > 0$ such that $v^{-1}(B(y_0, r)) \cap B(x_0, R) \subset\subset B(x_0, R)$.

**Proof** Define

$$d(\delta) := \sup\{|x - x_0|_2 : x \in \bar{B}(x_0, R), \ |v(x) - v(x_0)|_2 \leq \delta\}.$$

Since $v(x) \neq v(x_0)$ for every $x \in \bar{B}(x_0, R) \setminus \{x_0\}$ and $v$ is uniformly continuous on $\bar{B}(x_0, R)$, we have

$$\lim_{\delta \to 0} d(\delta) = 0.$$

Now take $r > 0$ such that $d(r) < \frac{R}{2}$. We have

$$v^{-1}(B(y_0, r)) \cap B(x_0, R) \subset B\left(x_0, \frac{R}{2}\right) \subset\subset B(x_0, R).$$

□

**Proof of Theorem 6.1** Let $\Omega'$ be the set of points $x_0 \in \Omega$ such that $v$ is completely differentiable at $x_0$ and $\det \nabla v(x_0) > 0$. By Theorem 5.21 we have $\mathcal{L}^N(\Omega \setminus \Omega') = 0$. In the sequel, we fix $x_0 \in \Omega'$, we set $y_0 = v(x_0)$, and we show that $v$ is locally invertible at $x_0$. By Lemmas 5.9 and 6.5 there exists $R_0 > 0$ such that $B(x_0, R_0) \subset\subset \Omega$,

$$N(v, B(x_0, R_0), y) = 1 \text{ a.e. } y \in C_{R_0}, \tag{6.11}$$

where $C_{R_0}$ is the connected component of $\mathbb{R}^N \setminus v(\partial B(x_0, R_0))$ containing $y_0$, with $N(v, B(x_0, R_0), y_0) = 1$. By Lemma 6.7 we deduce that there exists $r > 0$ such that

$$v^{-1}(B(y_0, r)) \cap B(x_0, R_0) \subset\subset B(x_0, R_0) \tag{6.12}$$

and

$$B(y_0, r) \subset\subset C_{R_0}. \tag{6.13}$$

Setting $D := v^{-1}(B(y_0, r)) \cap B(x_0, R_0)$, by (6.12) and (6.13) we have $D \subset v^{-1}(C_{R_0}) \cap B(x_0, R_0)$ and, by Lemma 6.6,

$$v(D) = B(y_0, r), \quad v(\partial D) \subset \partial v(D) = \partial B(y_0, r). \tag{6.14}$$

By the $N^{-1}$-property of $v$ (see Theorem 5.32) and (6.14), we have $\mathcal{L}^N(\partial D) = 0$ which, together with (6.6), yields

$$d(v, D, y) = 1 \quad \text{for all } y \in v(D) \setminus v(\partial D). \tag{6.15}$$

Using the definition of $D$, the fact that $D \subset B(x_0, R_0)$, (6.11), and (6.13), we obtain

$$N(v, D, y) = 1 \text{ a.e. } y \in v(D). \tag{6.16}$$

Let $E := \{y \in v(D) = B(y_0, r) : N(v, D, y) \neq 1\}$, so that $\mathcal{L}^N(E) = 0$. We define the candidate for the local inverse function, $w$, by

$$w(y) := x \text{ if } y \in v(D) \setminus E, \text{ and } v(x) = y, x \in D, \tag{6.17}$$
$$w(y) := x \text{ if } y \in E, \ v(x) = y, \tag{6.18}$$

$x \in D$ being chosen by the axiom of choice.
*Claim1.* $w \in L^\infty(B(y_0, r))^N$.

We have $w(y) \in D \subset \Omega$ for every $y \in v(D)$ and so $w$ is uniformly bounded in $v(D)$. To prove that $w$ is Lebesgue measurable, we fix $\alpha \in \mathbb{R}$ and show that the set

$$A := \{y \in v(D) : w_i(y) \geq \alpha\}$$

is measurable, $i = 1, \ldots, N$. Clearly, $A = A_1 \cup A_2$, where

$$A_1 := \{y \in v(D) \setminus E : w_i(y) \geq \alpha\}, \quad A_2 := \{y \in E : w_i(y) \geq \alpha\}.$$

Since $\mathcal{L}^N(A_2) = 0$ we deduce that $A_2$ is measurable. Using the fact that the restriction of $v$ to $v^{-1}(v(D) \setminus E)$ is one-to-one, one can see that

$$A_1 = \{v(x) : x \in v^{-1}(v(D) \setminus N), \ x_i \geq \alpha\}$$
$$= (v(D) \setminus E) \cap \left[ \bigcup_{n \in \mathbb{N}} v\left(\{x \in \bar{B}(x_0, R_0) : \alpha + n \leq x_i \leq \alpha + n + 1\}\right) \right].$$

Using the fact that for every $n \in \mathbb{N}$, $\{x \in \bar{B}(x_0, R_0) : \alpha + n \leq x_i \leq \alpha + n + 1\}$ is a compact set, $v$ is a continuous function, and $v(D) \setminus E$ is measurable we conclude that $A_1$ is measurable.
*Claim 2.*

$$v \circ w(y) = y \quad \text{for every } y \in v(D) = B(y_0, r), \tag{6.19}$$

$$w \circ v(x) = x \text{ for every } x \in D \setminus v^{-1}(E). \tag{6.20}$$

These follow immediately from (6.17) and (6.18). We notice that, due to Theorem 5.32,

$$\mathcal{L}^N(v^{-1}(N)) = 0.$$

*Claim 3.* $f \circ w$ is measurable for every $f : D \to \mathbb{R}$ measurable.

Since every Lebesgue measurable set is a union of a Borel measurable set and a set of measure zero, to show that $f \circ w$ is measurable, by Claim 1 it suffices to show that $w^{-1}(R)$ is measurable for every $R \subset D$ such that $\mathcal{L}^N(R) = 0$. Indeed, by (6.19),

$$w^{-1}(R) \subset v(R),$$

and by the $N$-property of $v$ we obtain that $\mathcal{L}^N(w^{-1}(R)) = 0$. Thus $w^{-1}(R)$ is measurable.

Let $g : v(D) = B(y_0, r) \to \mathbb{R}$ be defined by

$$g(y) := \frac{|\mathrm{adj}\nabla v(w(y))|_2}{\det \nabla v(w(y))}.$$

*Claim 4.* $g \in L^1(v(D))$.

By Claim 3 $g$ is measurable and by Theorem 5.23, Claim 2, and (6.16) we obtain

$$\int_{\mathbb{R}^N} N(v, D, y)\, |g(y)|\, dy = \int_{v(D)} |g(y)|\, dy$$

$$= \int_D |g \circ v(x)| \det \nabla v(x)\, dx = \int_D |\mathrm{adj}\nabla v(x)|_2\, dx.$$

Therefore, $g \in L^1(v(D))$.

*Claim 5.* $w \in W^{1,1}(v(D))^N$ and $\nabla w(y) = \left(\frac{\mathrm{adj}\nabla v(w(y))}{\det \nabla v(w(y))}\right)^T$.

To prove this claim, we fix $\phi \in C_0^\infty(v(D))$ and set $K := \mathrm{spt}\,\phi$. We show that

$$\int_{v(D)} w_\alpha(y) \frac{\partial \phi}{\partial y_j}(y)\, dy = -\int_{v(D)} \frac{\left(\mathrm{adj}\nabla v(w(y))\right)_\alpha^j}{\det \nabla v(w(y))} \phi(y)\, dy,$$

where $A_\alpha^j$ denotes the component of the $j$th row and the $\alpha$th column of a matrix $A$. Set $\delta := \mathrm{dist}(K, \partial v(D)) > 0$. Using the uniform continuity of $v$ on $\bar{D} \subset B(x_0, R_0)$ we choose $\epsilon > 0$ such that

$$|v(x) - v(x')|_2 \le \frac{\delta}{4} \text{ for every } x, x' \in \bar{D}, \ |x - x'|_2 \le \epsilon. \tag{6.21}$$

Let $\{v_n\} \subset C^\infty(\bar{D})^N$ be such that

$$v_n \to v \ \text{ in } \ C(\bar{D})^N \tag{6.22}$$

and

$$v_n \rightharpoonup v \ \text{ in } \ W^{1,N}(D)^N.$$

By (6.22) we can assume that

$$|v - v_n| \le \frac{\delta}{4} \ \text{ for every } \ n \in \mathbb{N}. \tag{6.23}$$

Since $v(\partial D) \subset \partial v(D)$ (see (6.14)), by (6.21) and (6.23) we have that

$$\text{dist}(x, \partial D) < \epsilon \ \text{ implies } \ \phi(v_n(x)) = 0$$

and so

$$\phi \circ v_n \in C_0^\infty(D).$$

By Theorem 5.23, (6.16), (6.20), and the fact that for every $n \in \mathbb{N}$ and for every $j = 1, \ldots, N$,

$$\sum_{\alpha=1}^N \frac{\partial}{\partial x_\alpha} (\text{adj}\nabla v_n)_\alpha^j = 0,$$

we have

$$
\begin{aligned}
\int_{v(D)} w_\alpha(y) \frac{\partial \phi}{\partial y_j}(y)\, dy &= \int_D w_\alpha(v(x)) \frac{\partial \phi}{\partial y_j}(v(x)) \det \nabla v(x)\, dx \\
&= \lim_{n \to +\infty} \int_D x_\alpha \frac{\partial \phi}{\partial y_j}(v_n(x)) \det \nabla v_n(x)\, dx \\
&= \lim_{n \to +\infty} \int_D x_\alpha \sum_{k=1}^N \left( \text{adj}\nabla v_n(x) \right)_k^j \frac{\partial}{\partial x_k} \phi(v_n(x))\, dx \\
&= -\lim_{n \to +\infty} \int_D \left( \text{adj}\nabla v_n(x) \right)_\alpha^j \phi(v_n(x))\, dx \\
&= -\int_D \left( \text{adj}\nabla v(x) \right)_\alpha^j \phi(v(x))\, dx \\
&= -\int_D \frac{\left( \text{adj}\nabla v(w \circ v(x)) \right)_\alpha^j}{\det \nabla v(w \circ v(x))} \phi(v(x)) \det \nabla v(x)\, dx \\
&= -\int_{v(D)} \frac{\left( \text{adj}\nabla v(w(y)) \right)_\alpha^j}{\det \nabla v(w(y))} \phi(y)\, dy.
\end{aligned}
$$

This equality, together with Claim 4, yields Claim 5.

*Claim 6.* $\nabla w \in W^{1,s}(v(D))$ if and only if $|g \circ v|^s \det \nabla v \in L^1(v(D))$ for $1 \le s < +\infty$.

Recall that $g(y) = \frac{|\mathrm{adj}\nabla v(w(y))|_2}{\det \nabla v(w(y))}$ and that $w \in L^\infty(v(D))^N$. Thus, we conclude that $\nabla w \in W^{1,s}(v(D))$ if and only if $\nabla w \in L^s(v(D))$. The result now follows from Claim 5 and Theorem 5.23. $\hfill \square$

**Remark 6.8** It is possible to show that if $v \in W^{1,q}(\Omega)^N$, $q > N - 1$, if $v$ is continuous, $\mathrm{adj}\nabla v \in L^{\frac{N}{N-1}}(\Omega)$, and $\det \nabla v(x) > 0$ a.e. in $\Omega$, then there is local invertibility a.e. in $\Omega$, i.e. for a.e. $x_0 \in \Omega$ there exists $r > 0$ such that $v|_{B(x_0,r)}$ is almost everywhere injective with the inverse $w \in BV_{loc}(v|_{v(B(x_0,r))})^N$ and there exists a set $E \subset v(B(x_0,r))$ such that

$$E \text{ is an open set of } v(B(x_0,r)),$$
$$\mathcal{L}^N(v(B(x_0,r) \setminus E)) = 0,$$
$$w \in W^{1,1}(E)^N,$$
$$v \circ w(y) = y \text{ a.e. } y \in v(B(x_0,r)),$$
$$w \circ v(x) = x \text{ a.e. } x \in B(x_0,r).$$

By Theorem 5.21 $v$ is weakly differentiable a.e. in $\Omega$ and by Lemma 5.10

$$d(v, B(x_0,r), v(x_0)) = 1$$

for some $r > 0$. Let $C_0$ be the connected component of $\mathbb{R}^N \setminus v(\partial B(x_0,r))$ which contains $v(x_0)$. Then

$$d(v, B(x_0,r), y) = 1 \tag{6.24}$$

for every $y \in C_0$ and so, if we choose $0 < r' < r$ such that

$$B(x_0,r') \subset B(x_0,r) \cap v^{-1}(C_0),$$

then, by (6.24) and since $\det \nabla v > 0$ a.e., repeating the argument used in (6.7)–(6.10) we have

$$d(v, B(x_0,r'), y) \leq 1$$

for every $y \in \mathbb{R}^N \setminus v(\partial B(x_0,r'))$. It suffices now to use Tang's results (1988, (1.3)–(1.5), (2.26), and Theorem 3.7 (i)). Note, however, that in Tang (1988) it is assumed that $\mathrm{adj}\nabla v \in L^r$, $r \geq \frac{q}{q-1}$ and if $N - 1 < q < N$, then $\frac{q}{q-1} > \frac{N}{N-1}$. As it turns out, Tang's results (1988) still hold for $r = \frac{N}{N-1}$, as was noted by Müller *et al.* (1994, Theorem 5.3).

**Proof of Corollary 6.2.**

(a) We have

$$v \in W^{1,N}(\Omega_1)^N, \quad \det \nabla v(x) > 0 \text{ a.e. } x \in \Omega_1$$

and

$$v \in W^{1,N}(\Omega_2)^N, \quad \det \nabla v(x) < 0 \text{ a.e. } x \in \Omega_2.$$

It suffices to apply Theorem 6.1 to $v$ in $\Omega_1$ and to $R_0 v$ in $\Omega_2$, where $R_0$ is a constant rotation with $\det R_0 = -1$.

(b) Now we assume that $v \in W^{1,q}(\Omega)^N$, $q > N$, $\det \nabla v(x) \neq 0$ a.e. $x \in \Omega$ and for almost every $x_0 \in \Omega$, $v$ is locally almost injective in a neighbourhood of $x_0$, in the sense that there is an open set $D \equiv D(x_0) \subset\subset \Omega$ and there is a function $w : v(D) \to D$ such that

$$w \circ v(x) = x \text{ a.e. } x \in D.$$

By a corollary of Vitali's Covering Theorem (see Corollary 4.35) there is a countable family of nonempty, open, mutually disjoint balls $\{B_i : i \in \mathbb{N}\}$ and there is a sequence of functions $w_i : v(\bar{B}_i) \to \Omega$ such that $\bar{B}_i \subset \Omega$ and

$$\mathcal{L}^N(\Omega \setminus \underset{i \in \mathbb{N}}{\cup} B_i) = 0,$$
$$w \circ v(x) = x \text{ a.e. } x \in B_i. \tag{6.25}$$

The task ahead will be to partition $B_i$ into three subsets $B_i^1, B_i^2$, and $N_i$ such that $B_i^1, B_i^2$ are two open sets and $N_i$ is a set of measure zero,

$$\det \nabla v(x) > 0 \text{ a.e. } x \in B_i^1,$$
$$\det \nabla v(x) < 0 \text{ a.e. } x \in B_i^2.$$

Using the fact that $v \in W^{1,q}(B_i)^N$, $q > N$, by (6.25) and Theorem 5.21 we deduce that there is a set of measure zero $A_i \subset \bar{B}_i$ such that $v$ is differentiable at every $x \in B_i \setminus A_i$,

$$w \circ v(x) = x \text{ for every } x \in \bar{B}_i \setminus A_i, \tag{6.26}$$
$$\det \nabla v(x) \neq 0 \text{ for every } x \in \bar{B}_i \setminus A_i.$$

Let $\{C^j\}$ be the countable collection of the (open) connected components of $\mathbb{R}^N \setminus v(\partial B_i)$. By Theorem 5.32 we have

$$\mathcal{L}^N(v^{-1}(v(\partial B_i \cup A_i))) = 0.$$

We claim the following.

*Claim 1.* $d(v, B_i, v(x)) = \operatorname{sgn} \det \nabla v(x)$ for every $x \in B_i \setminus v^{-1}(v(\partial B_i \cup A_i))$.

Fix $x \in B_i \setminus v^{-1}(v(\partial B_i \cup A_i))$. As $v$ is differentiable at $x$ and $\det \nabla v(x) \neq 0$, by Lemma 5.9 there exists $r_0 > 0$ such that for every $0 < r \leq r_0$ we have

$$d(v, B(x, r), v(x)) = \operatorname{sgn} \det \nabla v(x).$$

On the other hand, setting $K := \bar{B}_i \setminus B(x, r_0)$, $K$ is a compact set included in $\bar{B}_i$ and by (6.26) $v(x) \notin v(K)$ because $v(x) \notin v(A_i)$. By the excision property of the degree (see Theorem 2.7) we obtain

$$d(v, B_i, v(x)) = d(v, B(x, r_0), v(x)) = \operatorname{sgn} \det \nabla v(x).$$

*Claim 2.* sgn det $\nabla v(x) = $ sgn det $\nabla v(x')$ for every $x, x' \in v^{-1}(C^j) \setminus v^{-1}(v(\partial B_i \cup A_i))$.

Assume that $x, x' \in v^{-1}(C^j) \setminus v^{-1}(v(\partial B_i \cup A_i))$. Using Claim 1 and the fact that the degree $d(v, B_i, \cdot)$ is constant on each $C^j$, we obtain that sgn det $\nabla v(x) = $ sgn det $\nabla v(x')$.

Now we conclude the proof of (b). Let $I := \{j \in \mathbb{N} : \det \nabla v(x) > 0 \text{ a.e. } x \in v^{-1}(C^j)\}$ and $J := \{j \in \mathbb{N} : \det \nabla v(x) < 0 \text{ a.e. } x \in v^{-1}(C^j)\}$. Set

$$B_i^1 := \bigcup_{j \in I} v^{-1}(C^j) \cap B_i, \quad B_i^2 := \bigcup_{j \in J} v^{-1}(C^j) \cap B_i,$$

and

$$N_i := B_i \setminus (B_i^1 \cup B_i^2).$$

Then $B_i := B_i^1 \cup B_i^2 \cup N_i$ and by setting $\Omega_1 := \cup_i B_i^1$, $\Omega_2 := \cup_i B_i^2$ and $E := \Omega \setminus (\Omega_1 \cup \Omega_2)$, we obtain that $\mathcal{L}^N(E) = 0$ and $\Omega_1, \Omega_2$ have the required properties. □

**Proof of Corollary 6.3** To obtain that $w \in W^{1, \frac{q}{N-1}}(v(D), D)$ we take $s = \frac{q}{N-1}$ in Theorem 6.1. If $q \geq N(N-1)$, then $w \in W^{1,N}$,

$$\det \nabla w(y) = \frac{1}{\det \nabla v(w(y))} > 0 \quad \text{a.e. } y \in v(D)$$

and so, by Corollary 5.19, we deduce that $w$ is continuous. Hence $v$ and $w$ are homeomorphisms and $v$ is an open mapping in $\Omega'$ for some $\Omega' \subset \Omega$ open, where $\mathcal{L}^N(\Omega \setminus \Omega') = 0$. □

## 6.2    Energy functionals involving variation of the domain

The variational treatment of crystals with defects leads to the study of functionals of the type

$$E(u, v) := \int_\Omega W(\nabla u(x)(\nabla v(x))^{-1}) \, dx,$$

where $\Omega \subset \mathbb{R}^N$ is a reference domain, $W$ is the strain energy density, $u$ is the elastic deformation, and $v$ represents the slip (rearrangement) or plastic deformation, with $\det(\nabla v(x)) = 1$ a.e. $x \in \Omega$. The underlying kinematical model for slightly defective crystals was introduced by Davini (1986) and later developed by Davini and Parry (1989). As it turns out, matrices of the form

$$\nabla u(x)(\nabla v(x))^{-1}$$

represent lattice matrices of defect-preserving deformations (neutral deformations), and, taking the viewpoint that equilibria correspond to a variational principle, Fonseca and Parry (1987) studied the structure of some kinds of generalized minimizer (Young measure solutions) for the energy $E(\cdot, \cdot)$ ( related variational problems were also investigated in Fonseca and Parry 1987).

The lower semicontinuity and relaxation properties of $E(\cdot, \cdot)$ were addressed only under additional material symmetry assumptions on $W$. The existence and regularity properties for minimizers of $E(\cdot, \cdot)$ were obtained by Dacorogna and Fonseca (1992). Following this work, we stress the fact that the direct methods of the calculus of variations fail to apply to this problem, as sequential weak lower semicontinuity of $E(\cdot, \cdot)$ is not sufficient to guarantee the existence of minimizers. Indeed, if $W(F) = |F|_2^r$, it was shown by Dacorogna and Fonseca (1992) that there are no minimizers in $\{(u, v) \in W^{1,\infty} \times W^{1,\infty}| \ u(x) = x$ on $\partial\Omega, \ \det(\nabla v(x)) = 1$ a.e.$\}$ if $0 < r < N = 2$, while for $r > N$ existence is obtained for smooth $(u, v)$ (Dacorogna and Fonseca 1992, Theorem 2.3).

It is clear that if $\{(u_n, v_n)\}$ is a minimizing sequence and if $|\nabla u_n(\nabla v_n)^{-1}|_2^r$ is bounded in $L^1$, then

$$\nabla u_n(\nabla v_n)^{-1} \rightharpoonup L \text{ in } L^r, u_n|_{\partial\Omega} = u_0, \ \det(\nabla v_n) = 1 \text{ a.e.}$$

and so, if some type of lower semicontinuity prevails, then

$$\int_\Omega W(L)\, dx \leq \liminf_{n \to +\infty} \int_\Omega W(\nabla u_n(\nabla v_n)^{-1})\, dx. \tag{6.27}$$

It would remain to show that $L$ will still have the same structure, precisely

$$L = \nabla u(\nabla v)^{-1}$$

where $u|_{\partial\Omega} = u_0$, $\det(\nabla v) = 1$ a.e. $x \in \Omega$. Using the div–curl lemma it follows that if $u_n \rightharpoonup u$ in $W^{1,\infty}$ $w-*$ and $v_n \rightharpoonup v$ in $W^{1,\infty}$ $w-*$, then

$$\nabla u_n(\nabla v_n)^{-1} \rightharpoonup \nabla u(\nabla v)^{-1} \text{ in } L^\infty \ w-*.$$

Note that (6.27) is always satisfied if $W$ is a convex function. On the other hand, formally, as $\det(\nabla v) = 1$ a.e. and setting $w = u(v^{-1})$, the energy becomes

$$\int_{v(\Omega)} W(\nabla w(y))\, dy,$$

which is now an energy functional involving variations of the domain. Hence, under this new formulation, quasiconvexity seems to be more appropriate than convexity (see Acerbi and Fusco 1983, Ball 1978, and Dacorogna 1987).

Suppose that $W$ is a *quasiconvex function*, i.e.

$$W(F) \leq \frac{1}{\mathcal{L}^N(Q)} \int_Q W(F + \nabla\phi(x))\, dx,$$

where $Q = (0, 1)^N, \phi \in W_0^{1,\infty}(Q)^N$, and let $\nabla u_n(\nabla v_n)^{-1} \rightharpoonup L$ in $L^r$. Can we say that

$$\int_\Omega W(L) \leq \liminf_{n \to +\infty} \int_\Omega W(\nabla u_n(\nabla v_n)^{-1})?$$

As an example, consider

$$W(F) := |F|_2^2 + |\det(F)|.$$

Although we are unable to answer this question, we prove the following related result.

**Theorem 6.9** *Let $W : M^{N \times N} \to \mathbb{R}$ be a quasiconvex function such that*

$$-C_1(1 + |A|_2^s) \leq W(A) \leq C_2(1 + |A|_2^r)$$

*for some constants $C_1, C_2 > 0, r > s \geq 1$, $p \geq 1$, $q \geq N$, $\frac{1}{p} + \frac{N-1}{q} = \frac{1}{r}$ ($W \geq 0$ if $r = s = 1$). If $u_n \rightharpoonup u$ in $W^{1,p}(\Omega)^N$, $v_n \rightharpoonup v$ in $W^{1,q}(\Omega)^N$, and if $\det(\nabla v_n) = 1$ a.e. in $\Omega$, then*

$$\int_\Omega W(\nabla u (\nabla v)^{-1}) \, dx \leq \liminf_{n \to +\infty} \int_\Omega W(\nabla u_n (\nabla v_n)^{-1}) \, dx.$$

Before proving Theorem 6.9 we make some remarks.

**Remark 6.10**

(i) It is clear that if $u \in W^{1,p}$, $v \in W^{1,q}$ and $\det \nabla v = 1$ a.e., then $\nabla u (\nabla v)^{-1} \in L^r$.

(ii) If $r > 1$, then $s < r$ is a necessary condition, as the counterexample by Murat and Tartar shows (see Ball and Murat 1984). Here $r = s = 2 = N$, $\Omega = (0,1)^2$, $W(F) = \det(F)$, $u_n \rightharpoonup u$ in $H^1(\Omega)$, $v_n(x) = x$, and

$$\int_\Omega \det \nabla u \not\leq \liminf \int_\Omega \det \nabla u_n.$$

(iii) The growth condition cannot be dropped even if $W$ is polyconvex and nonnegative. Precisely, if the relation between $p, q, r$, and $s$ does not hold, the conclusion of Theorem 6.9 may be false. Indeed, using the example by Malý (1993) with $q = +\infty, p < N - 1, W(F) = \det F, N = r = s$, we may find $u_n \rightharpoonup u$ in $W^{1,p}$, $u(x) \equiv x$ with $v_n(x) \equiv x$, and

$$\int_\Omega |\det(\nabla u)| > \liminf \int_\Omega |\det(\nabla u_n)|.$$

Moreover, the growth condition prescribed in Theorem 6.9 is the well-known growth condition ensuring weak lower semicontinuity of $E(u, id)$ in $W^{1,p}$ (see Acerbi and Fusco 1983 and Dacorogna 1987).

(iv) It is natural to ask whether or not these results can be extended to the case $\frac{N^2}{N+1} < q < N$, since, due to Müller's result (1990b) and if we assume that $\det \nabla v = 1$ a.e., then $\mathrm{Det}\, \nabla v = \det \nabla v$ a.e. in $\Omega$.

(v) Having obtained lower semicontinuity of the energy in Theorem 6.9, the question now amounts to showing that one can find a minimizing sequence $\{\nabla u_n (\nabla v_n)^{-1}\}$ where $\{u_n\}$ is bounded in $W^{1,p}$ and $\{v_n\}$ is bounded in $W^{1,\infty}$. Actually, one only needs to show that there exists a sequence $\{f_n\} \subset W^{1,\infty}(\Omega, \Omega)$ such that $v_n \circ f_n$ is bounded in $W^{1,\infty}$ and

$$\begin{cases} \det \nabla f_n(x) = 1 & \text{a.e. } x \in \Omega \\ f_n(x) \quad\;\; = 1 & x \in \partial\Omega, \end{cases}$$

since $\nabla u_n (\nabla v_n)^{-1} = \nabla(u_n \circ f_n) \cdot (\nabla(v_n \circ f_n))^{-1}$. Due to the examples provided by Dacorogna and Fonseca (1992), we know that this may not be possible since the infimum of $E$ may be zero, which prevents the existence of a minimizing sequence bounded in $W^{1,p} \times W^{1,q}$.

As is usual in variational problems for which existence of minimizers is not guaranteed (such as variational problems for material that changes phase or, as in our case, for slightly defective materials), rather then studying the macroscopic limit of $\nabla u_n (\nabla v_n)^{-1}$ we focus our attention on the properties of the minimizing sequences.

The following may contribute to a better understanding of why the boundedness of $\{\nabla u_n (\nabla v_n)^{-1}\}$ may not entail the boundedness of $\{\nabla u_n\}$ and $\{\nabla v_n\}$. Using Theorem 6.9 we show that we can construct a minimizing sequence of the form $\{\nabla u_\epsilon (\nabla v_\epsilon)^{-1}\}$ with $||\nabla u_\epsilon||_p = 0(\frac{1}{\epsilon^\alpha})$, $||\nabla v_\epsilon||_q = 0(\frac{1}{\epsilon^\beta})$, for any $\alpha, \beta > 0$.

Consider the 'perturbed' family of variational problems

$$E_\epsilon(u, v) := \int_\Omega W(\nabla u(\nabla v)^{-1}) \, dx + \epsilon^{\alpha p}||\nabla u_\epsilon||_p^p + \epsilon^{\beta q}||\nabla v_\epsilon||_q^q,$$

where $u|_{\partial\Omega} = u_0$, $\det \nabla v = 1$ a.e., and $\frac{1}{\mathcal{L}^N(\Omega)} \int_\Omega v(x) \, dx = 0$. Using the direct method of the Calculus of Variations, Poincaré's Inequality and Theorem 6.9, it follows immediately that there exists $(u_\epsilon, v_\epsilon) \in W^{1,p} \times W^{1,q}$ such that

$$E_\epsilon(u_\epsilon, v_\epsilon) = \inf \left\{ E_\epsilon(u, v) : \ (u, v) \in W^{1,p} \times W^{1,q}, \ \det \nabla v = 1 \ \text{a.e.} \right\}.$$

Then, given an admissible pair $(u, v)$,

$$\begin{aligned} E(u, v) &= \lim_{\epsilon \to 0+} E_\epsilon(u, v) \\ &\geq \limsup_{\epsilon \to 0+} E_\epsilon(u_\epsilon, v_\epsilon) \\ &\geq \limsup_{\epsilon \to 0+} E(u_\epsilon, v_\epsilon), \\ &\geq \inf E. \end{aligned}$$

Using the same reasoning with $\liminf_{\epsilon \to 0+} E(u_\epsilon, v_\epsilon)$ and taking the infimum in $(u, v)$ we conclude that

$$\inf E = \lim_{\epsilon \to 0+} E(u_\epsilon, v_\epsilon)$$

and

$$||\nabla u_\epsilon||_p = 0(\frac{1}{\epsilon^\alpha}), \quad ||\nabla v_\epsilon||_q = 0(\frac{1}{\epsilon^\beta}).$$

The following two lemmas will be useful in proving Theorem 6.9.

**Lemma 6.11** *Let $\Omega', \Omega$ be two open sets of $\mathbb{R}^N$ such that $\Omega' \subset\subset \Omega$, let $q \geq N$, and let $v, v_n \in W^{1,q}(\Omega)^N$ be such that $\det \nabla v(x) = \det \nabla v_n(x) = 1$ a.e. $x \in \Omega$. Assume that $v_n \rightharpoonup v$ in $W^{1,q}(\Omega)^N$. Then there exists a subsequence of $\{v_n\}$ (not relabelled) such that for almost every $x_0 \in \Omega'$ there exist open sets $D, D_n \subset \Omega'$ containing $x_0$ and there exist $n_0 \in \mathbb{N}$, $r_0 \equiv r(x_0) > 0$, $w : B(y_0, r_0) \to D$, $w_n : B(y_0, r_0) \to D_n$, with $y_0 = v(x_0)$ such that for $n \geq n_0$,*

$$w_n \circ v_n(x) = x \quad \text{a.e. } x \in \bar{D}_n,$$
$$v_n \circ w_n(y) = y \quad \text{for every } y \in \bar{B}(y_0, r_0) \quad \text{and} \quad v_n(D_n) = B(y_0, r_0),$$
$$w \circ v(x) = x \quad \text{a.e. } x \in \bar{D} \quad \text{and} \quad v(x_0) \neq v(x) \text{ for } x \in D, x \neq x_0,$$
$$v \circ w(y) = y \quad \text{for every } y \in \bar{B}(y_0, r_0) \quad \text{and} \quad v(D) = B(y_0, r_0),$$
$$w_n, w \in W^{1, \frac{q}{N-1}}.$$

**Proof** Using Corollary 5.19 and the Ascoli–Arzela Theorem we obtain that, up to a subsequence, $v_n$ converges to $v$ uniformly in $\bar{\Omega}'$. By Lemma 6.5 for almost every $x_0 \in \Omega'$, there is $R_0 > 0$ such that

$$B(x_0, R_0) \subset\subset \Omega',$$
$$N(v, B(x_0, R_0), y) = 1 \text{ for almost every } y \in C_{R_0},$$
$$d(v, B(x_0, R_0), y) = 1 \text{ for every } y \in C_{R_0},$$
$$d(v, B, y) = 1 \text{ for every } y \in B \setminus v(\partial B),$$

for every nonempty open set $B \subset v^{-1}(C_{R_0}) \cap B(x_0, R_0)$ such that $\mathcal{L}^N(v(\partial B)) = 0$, where $C_{R_0}$ is the connected component of $\mathbb{R}^N \setminus v(\partial B(x_0, R_0))$ containing $y_0 := v(x_0)$. Since $v$ is differentiable at $x_0$ and $\det \nabla v(x_0) \neq 0$ we may assume without loss of generality that $N(v, B(x_0, R_0), y_0) = 1$. Fix $0 < \epsilon < d(y_0, v(\partial B(x_0, R_0)))$ and choose $n_0 \in \mathbb{N}$ such that $||v_n - v|| < \epsilon$. Set

$$A_\epsilon := \{y \in C_{R_0} : \text{dist}(y, v(\partial B(x_0, R_0))) > \epsilon\}.$$

It is obvious that $A_\epsilon$ is a nonempty open set.
*Claim 1.* $d(v_n, B(x_0, R_0), y)$ exists and is equal to 1 for every $y \in A_\epsilon$ and every $n \geq n_0$.

By Theorem 2.3, together with the fact that $d(v, B(x_0, R_0), y) = 1$ for every $y \in C_{R_0}$, we have

$$d(v_n, B(x_0, R_0), y) = 1 \tag{6.28}$$

for every $y \in A_\epsilon$ and every $n \geq n_0$. By Lemma 6.7 there is $0 < r_0 < R_0$ such that

$$B(y_0, r_0) \subset\subset A_\epsilon \quad \text{and} \quad v^{-1}(B(y_0, r_0)) \cap B(x_0, R_0) \subset\subset B(x_0, R_0). \tag{6.29}$$

*Claim 2.* We claim that

$$B(y_0, r_0) \subset\subset C^n_{R_0} \tag{6.30}$$

where $C^n_{R_0}$ is the connected component of $\mathbb{R}^N \setminus v_n(\partial B(x_0, R_0))$ containing $y_0$.

We first prove that $A_\epsilon \subset \mathbb{R}^N \setminus v_n(\partial B(x_0, R_0))$. Assume, on the contrary, that there is $y \in A_\epsilon \cap v_n(\partial B(x_0, R_0))$ and choose $x \in \partial B(x_0, R_0)$ such that $y = v_n(x)$. We would then have $|v_n(x) - v(x)|_2 = |y - v(x)|_2 > \epsilon > ||v_n - v||$, which yields a contradiction. Fix $r' > r_0$ such that $\bar{B}(y_0, r') \subset A_\epsilon$. We have that $B(y_0, r')$ is a connected set included in $\mathbb{R}^N \setminus v_n(\partial B(x_0, R_0))$ and containing $y_0$. We deduce that $B(y_0, r') \subset C^n_{R_0}$ and $B(y_0, r_0) \subset\subset C^n_{R_0}$.

Set

$$D := v^{-1}(B(y_0, r_0)) \cap B(x_0, R_0) \subset\subset \Omega',$$
$$D_n := v_n^{-1}(B(y_0, r_0)) \cap B(x_0, R_0) \subset\subset \Omega'.$$

By (6.28), (6.29), (6.30), and using arguments similar to those in the proof of Theorem 6.1, together with Corollary 6.3, we deduce that for $n \geq n_0$ there exists $w_n : \bar{B}(y_0, r_0) \to \bar{D}_n$ and $w : \bar{B}(y_0, r_0) \to \bar{D}$ such that

$$w_n, w \in W^{1, \frac{q}{N-1}}(B(y_0, r_0))^N,$$

and

$$w_n \circ v_n(x) = x \quad \text{a.e. } x \in \bar{D}_n,$$
$$v_n \circ w_n(y) = y \quad \text{a.e. } y \in \bar{B}(y_0, r_0),$$
$$w \circ v(x) = x \quad \text{a.e. } x \in \bar{D} \text{ and } v(x_0) \neq v(x) \text{ for } x \in \bar{D}, x \neq x_0,$$
$$v \circ w(y) = y \quad \text{a.e. } y \in \bar{B}(y_0, r_0).$$

Finally, by Lemma 6.6, we conclude that $v_n(D_n) = v(D) = B(y_0, r_0)$. □

**Remark 6.12**

(i) It can easily be seen by the preceding proof that if the conclusion of Lemma 6.11 holds for $r \equiv r(x_0) > 0$, then it also holds for $0 < r' < r$. Thus, as $v$ is continuous on $\bar{D}$, $v(x) \neq v(x_0)$ for $x \in D$ and $x \neq x_0$, we deduce that

$$\lim_{r \to 0} \max\{|x - x_0|_2 : x \in \bar{D}, v(x) \in B(y_0, r)\} = 0.$$

(ii) It is possible to show that $\lim_{n \to +\infty} \mathcal{L}^N(D \Delta D_n) = 0$. First, we prove that $\lim_{n \to +\infty} \mathcal{L}^N(D \setminus D_n) = 0$.

Let $F_\epsilon := B(y_0, r_0 - \epsilon)$ and $O_\epsilon := v^{-1}(F_\epsilon) \cap D$. We claim that for each fixed $\epsilon$ there exists $n_0 \equiv n_0(\epsilon) \in \mathbb{N}$ such that $n \geq n_0$ implies $O_\epsilon \subset D_n$. Indeed, since $\{v_n\}$ converges to $v$ uniformly, there exists $n_0 \equiv n_0(\epsilon) \in \mathbb{N}$ such that $||v - v_n|| \leq \frac{\epsilon}{2}$ for every $n \geq n_0$. If $x \in O_\epsilon$, we obtain

$$|v_n(x) - y_0|_2 \leq |v(x) - y_0|_2 + |v(x) - v_n(x)|_2 < r_0$$

and so $x \in D_n$. As $\cup_\epsilon O_\epsilon = D$ and the sequence $\{O_\epsilon\}$ is nonincreasing, we have

$$\lim_{\epsilon \to 0} \mathcal{L}^N(D \setminus O_\epsilon) = 0,$$

which, together with the fact that $\mathcal{L}^N(D \setminus D_n) \leq \mathcal{L}^N(D \setminus O_\epsilon)$ for $n \geq n_0$, yields $\lim_{n \to +\infty} \mathcal{L}^N(D \setminus D_n) = 0$. Next, we prove that $\lim_{n \to +\infty} \mathcal{L}^N(D_n \setminus D) = 0$. For $\epsilon > 0$ take $n_0 \equiv n_0(\epsilon) \in \mathbb{N}$ such that $||v - v_n|| \leq \frac{\epsilon}{2}$ for every $n \geq n_0$. We have

$$\{x \in B(x_0, R_0) : r - \tfrac{\epsilon}{2} \leq |v_n(x) - y_0|_2 < r\}$$
$$\subset \{x \in B(x_0, R_0) : r - \epsilon \leq |v(x) - y_0|_2 < r + \epsilon\}$$

and since $v$ has the $N^{-1}$-property (see Theorem 5.32), we obtain

$$\mathcal{L}^N(\bigcap_{\epsilon > 0} \{x \in B(x_0, R_0) : r - \epsilon \leq |v(x) - y_0|_2 < r + \epsilon\})$$
$$= \mathcal{L}^N(\{x \in B(x_0, R_0) : |v(x) - y_0|_2 = r\}) = 0.$$

To conclude it suffices to remark that for $n \geq n_0$ we obtain

$$D_n \setminus D \subset \{x \in B(x_0, R_0) : r - \epsilon \leq |v(x) - y_0|_2 < r + \epsilon\}.$$

**Lemma 6.13** *Let $p \geq 1$, $q \geq N$, $r \geq 1$ be such that $\frac{1}{p} + \frac{N-1}{q} = \frac{1}{r}$. Let $\Omega \subset \mathbb{R}^N$ be an open, bounded set, $u_n, u \in W^{1,q}(\Omega)^N$, $u_n \rightharpoonup u$ in $W^{1,p}(\Omega)^N$, $v_n, v \in W^{1,q}(\Omega)^N$, $\det \nabla v_n = \det \nabla v = 1$ a.e. in $\Omega$ and let $v_n \rightharpoonup v$ in $W^{1,q}(\Omega)^N$. Let $x_0 \in \Omega$, let $w_n$, $w$ be, respectively, the local inverse function of $v_n$, $v$, in the open neighbourhoods $D_n$, $D$ of $x_0$, and let $y_0 = v(x_0)$ and $B(y_0, r_0)$ be as in Lemma 6.11 and Remark 6.12. Then the following hold:*
  (i) $u_n \circ w_n \in W^{1,r}(B(y_0, r_0))^N$, $\nabla(u_n \circ w_n)(y) = \nabla u_n(w_n(y))(\nabla v_n(w_n(y)))^{-1}$ *for almost every $y \in v_n(D_n)$,*
  (ii) $u_n \circ w_n \rightharpoonup u \circ w$ in $W^{1,r}(B(y_0, r_0))^N$ *if $r > 1$,*
  (iii) $u_n \circ w_n \to u \circ w$ *in $L^1(B(y_0, r_0))^N$ and the sequence $\{u_n \circ w_n\}$ is bounded in $W^{1,1}(B(y_0, r_0))^N$ if $r = 1$.*

**Proof** We recall that by Lemma 6.11 we have

$$w_n, w \in W^{1, \frac{q}{N-1}}(B(y_0, r_0))^N, v(D) = B(y_0, r_0), v_n(D_n) = B(y_0, r_0), \quad (6.31)$$
$$\nabla w(y) = (\nabla v(w(y)))^{-1}, \nabla w_n(y) = (\nabla v_n(w_n(y)))^{-1} \text{a.e. } y \in B(y_0, r_0), \quad (6.32)$$
$$N(v, D, y) = N(v_n, D_n, y) = 1 \text{ a.e. } y \in B(y_0, r_0), \quad (6.33)$$
$$w \circ v(x) = x \text{ a.e. } x \in D, w_n \circ v_n(x) = x \quad \text{a.e. } x \in D_n. \quad (6.34)$$

*First step.* We prove that $u \circ w, u_n \circ w_n \in W^{1,r}(B(y_0, r_0))^N$.

In fact, by the change of variables formula (see Theorem 5.23), (6.31), (6.32), (6.33), and (6.34) we have

$$\int_{B(y_0,r_0)} |u \circ w(y)|_2^r \, dy = \int_{v(D)} |u \circ w(y)|_2^r \, N(v, D, y) \, dy$$

$$= \int_D |u(x)|_2^r \, dx < +\infty$$

and so

$$u \circ w, \ u_n \circ w_n \in L^r(B(y_0, r_0))^N.$$

Let $\phi \in C_0^\infty(B(y_0, r_0))$. By Theorem 5.23, (6.31), (6.32), (6.33), (6.34), and using the fact that each vector row of adj $\nabla v$ is divergence free, we have

$$\int_{B(y_0,r_0)} u_i \circ w(y) \frac{\partial \phi}{\partial y_j} \, dy = \int_D u_i(x) \frac{\partial \phi}{\partial y_j} \circ v(x) \, dx$$

$$= -\int_D \sum_{l=1}^N \frac{\partial u_i}{\partial x_l}(x)((\nabla v(x))^{-1})_j^l \phi \circ v(x) \, dx$$

$$= -\int_{B(y_0,r_0)} \sum_{l=1}^N \frac{\partial u_i}{\partial x_l}(w(y))((\nabla v)^{-1} \circ w(y))_j^l \phi(y) \, dy.$$

Thus,

$$u \circ w \in W^{1,r}(B(y_0, r_0))^N$$

and

$$\nabla u \circ w(y) = \nabla u(w(y))(\nabla v(w(y)))^{-1} \quad \text{a.e. in} \quad B(y_0, r_0).$$

We have a similar result for $u_n \circ w_n$.

*Second step.* We conclude that $\{u_n \circ w_n\}$ is bounded in $W^{1,r}(B(y_0, r_0))^N$.
Indeed,

$$\int_{B(y_0,r_0)} |u_n \circ w_n(y)|_2^r \, dy = \int_{D_n} |u_n(x)|_2^r \, dx \le \int_\Omega |u_n(x)|_2^r \, dx$$

and since $r \le p$ and $\{u_n\}$ is bounded in $W^{1,p}(\Omega)^N$, we deduce that $\{u_n \circ w_n\}$ is bounded in $L^r(B(y_0, r_0))^N$. Also,

$$\int_{B(y_0,r_0)} |\nabla u_n \circ w_n(y)|_2^r \, dy = \int_D |\nabla u_n(x)(\nabla w_n(x))^{-1}|_2^r \, dx$$

$$\le C' \left[ \int_\Omega |\nabla u_n(x)|_2^p \, dx \right]^{\frac{r}{p}} \left[ \int_\Omega |\nabla v_n(x)|_2^{\frac{q}{N-1}} \, dx \right]^{\frac{r(N-1)}{p}}$$

$$\le C$$

for some constant $C$ which does not depend on $y_0, r$, and $n$. Thus $\{u_n \circ w_n\}$ is bounded in $W^{1,r}(B(y_0, r_0))^N$.

*Third step.* We prove that, up to a subsequence, $\{u_n \circ w_n\}$ converges strongly in $L^1(B(y_0, r_0))$ to $u \circ w$.

Let $f \in C(\bar{B}(y_0, r_0))$. By virtue of Remark 6.12, $\lim_{n \to +\infty} \mathcal{L}^N(D \Delta D_n) = 0$ and so

$$\chi_{D_n}(x) \to \chi_D(x) \quad \text{a.e. } x \in \Omega.$$

Using the fact that $u_n \rightharpoonup u$ in $W^{1,p}(\Omega)^N$, $v_n \rightharpoonup v$ in $W^{1,q}(\Omega)^N$ and assuming, without loss of generality, that $u_n \to u$ a.e., $v_n \to v$ a.e., by Theorem 5.23 and the Lebesgue Dominated Convergence Theorem, we obtain that

$$\lim_{n \to +\infty} \int_{B(y_0, r_0)} u_n \circ w_n(y) f(y) \, dy = \lim_{n \to +\infty} \int_{D_n} u_n(x) f(v_n(x)) \, dx$$

$$= \int_D u(x) f(v(x)) \, dx$$

$$= \int_{B(y_0, r_0)} u \circ w(y) f(y) \, dy.$$

Therefore, $u_n \circ w_n$ converges strongly to $u \circ w$ in measure and applying the Sobolev Imbedding Theorem to the bounded sequence $\{u_n \circ w_n\}$ in $W^{1,r}(\Omega)$, we conclude that, up to a subsequence, $u_n \circ w_n$ converges strongly in $L^1(B(y_0, r_0))$ to $u \circ w$.

*Fourth step.* Using the second and the third step we have that $\{\nabla u_n \circ w_n\}$ is bounded in $W^{1,r}(\Omega)^N$,

$$u_n \circ w_n \rightharpoonup u \circ w \quad \text{in } W^{1,r}(\Omega)^N \quad \text{if } r > 1$$

and

$$u_n \circ w_n \to u \circ w \quad \text{in } L^1(\Omega)^N \quad \text{if } r = 1.$$

$\square$

Now we give the proof of Theorem 6.9.

**Proof of Theorem 6.9.** Without loss of generality (and, if necessary, after extracting a subsequence of $\{(u_n, v_n)\}$), we assume that

$$\liminf_{n \to +\infty} \int_\Omega W(\nabla u_n(x)(\nabla v_n(x))^{-1}) \, dx = \lim_{n \to +\infty} \int_\Omega W(\nabla u_n(x)(\nabla v_n(x))^{-1}) \, dx < +\infty.$$

Fix $\epsilon > 0$ and let $\Omega_\epsilon \subset\subset \Omega$ be an open set such that $\mathcal{L}^N(\Omega \setminus \Omega_\epsilon) < \epsilon$. By Corollary 5.19 and the Ascoli–Arzela Theorem, without loss of generality we may assume that $v_n$ converges to $v$ uniformly in $\bar{\Omega}_\epsilon$. Set

$$C := \{x \in \Omega_\epsilon : v \text{ is differentiable and almost invertible at } x\},$$

$$A := \{D(x) : x \in C, D(x) \text{ is an open set of } \Omega_\epsilon, v(D(x)) \text{ is an open ball}\},$$

and

$$\Omega'_\epsilon := \bigcup_{D \in A} D.$$

As in the proof of Lemma 6.7, it is easy to see that

$$\inf\{\text{diam } D(x) \; : \; D(x) \in A\} = 0,$$

for every $x \in C$. By Lemma 6.11 and by a corollary of Vitali's Covering Theorem (see Corollary 4.35) there exist $\{x^j : j \in \mathbb{N}\} \subset \Omega_\epsilon$, $\{D^j : j \in \mathbb{N}\}$ a family of mutually disjoint, open neighbourhoods of, respectively, $x_j$, $N$ a set of measure zero such that,

$$\Omega_\epsilon = N \underset{j \in \mathbb{N}}{\cup} D^j,$$

and $v : D^j \to B(y^j, r^j)$ admits an inverse $w^j \in W^{1, \frac{q}{N-1}}(B(y^j, r^j), D^j)$, in the sense of Theorem 6.1, for some $r_j > 0$ and with $y^j = v(x^j)$. Recall that

$$\nabla(u \circ w^j)(y) = \nabla u(w^j(y))(\nabla v(w^j(y)))^{-1} \;\; \text{a.e.} \;\; y \in B(y^j, r^j),$$
$$w^j \circ v(x) = x \;\; \text{a.e.} \;\; x \in D^j,$$
$$v \circ w^j(y) = y \;\; \text{a.e.} \;\; y \in B(y^j, r^j)$$

and $D^j = v^{-1}(B(y^j, r^j)) \cap B(x^j, R^j)$ for some $R^j > 0$. Fix $k \in \mathbb{N}$. By Lemma 6.13 we obtain for each $j = 1, \ldots, k$ and up to a subsequence, the existence of $w_n^j \in W^{1, \frac{q}{N-1}}(B(y^j, r^j))^N$, which is the inverse function of $v_n|_{D_n^j}$, where $D_n^j = v_n^{-1}(B(y^j, r^j)) \cap B(x^j, R^j)$. Recall that $\frac{1}{r} = \frac{1}{p} + \frac{N-1}{q}$ and also

$$w_n^j \circ v_n(x) = x \;\; \text{a.e.} \;\; x \in D_n^j,$$
$$u_n \circ w_n^j \in W^{1,r}(B(y^j, r^j))^N,$$
$$\nabla(u_n \circ w_n^j)(y) = \nabla u_n(w_n^j(y))(\nabla v_n(w_n^j(y)))^{-1} \;\; \text{a.e.},$$
$$u_n \circ w_n^j \to u \circ w^j \;\; \text{in} \;\; W^{1,r}(B(y^j, r^j))^N \;\; \text{if} \;\; r > 1,$$
$$u_n \circ w_n^j \to u \circ w^j \;\; \text{in} \;\; L^1(B(y^j, r^j))^N,$$
$$\{u_n \circ w_n^j\} \;\; \text{is bounded in} \;\; W^{1,1}(B(y^j, r^j))^N \;\; \text{if} \;\; r = 1,$$
$$\lim_{n \to +\infty} \mathcal{L}^N(D_n^j \Delta D^j) = 0.$$

Fix

$$0 < \eta < \min\{r^j : j = 1, \ldots, k\}.$$

There exists $n(\eta) \in \mathbb{N}$ such that for every $n \geq n(\eta)$ we obtain

$$\max\{|v_n(x) - v(x)|_2 \; : \; x \in \Omega_\epsilon\} < \eta.$$

Since $D^j = v^{-1}(B(y^j, r^j)) \cap B(x^j, R^j)$, we deduce that for every $n \geq n(\eta)$

$$D_n^j(\eta) := D_n^j \cap v_n^{-1}(B(y^j, r^j - \eta)) \subset D^j$$

and so $D_n^i \cap D_n^j = \emptyset$ if $i \neq j$. Set

$$D^j(\eta) := D^j \cap v^{-1}(B(y^j, r^j - \eta)).$$

We divide the rest of the proof of Theorem 6.9 into two cases.

*Case 1.* We assume that $1 = r = \frac{1}{p} + \frac{N-1}{q}$ and that there is a constant $C$ such that $0 \leq W(F) \leq C(1 + |F|_2)$ for every $F \in M^{N \times N}$. Since $W \geq 0$ and $\{D^j(\eta)\}, \{D_n^j(\eta)\}$ are mutually disjoint for every $n \in \mathbb{N}$, by Fonseca and Müller (1992) we have

$$\int_{\overset{k}{\underset{j=1}{\cup}} D^j(\eta)} W(\nabla u(x)(\nabla v)^{-1}(x)) \, dx = \sum_{j=1}^{k} \int_{D^j(\eta)} W(\nabla u(x)(\nabla v)^{-1}(x)) \, dx$$

$$= \sum_{j=1}^{k} \int_{B(y^j, r^j - \eta)} W((\nabla u \circ w^j)(y)) \, dy$$

$$\leq \sum_{j=1}^{k} \liminf_{n \to +\infty} \int_{B(y^j, r^j - \eta)} W((\nabla u_n \circ w_n^j)(y)) \, dy$$

$$= \sum_{j=1}^{k} \liminf_{n \to +\infty} \int_{D_n^j(\eta)} W(\nabla u_n(x)(\nabla v_n)^{-1}(x)) \, dx$$

$$\leq \liminf_{n \to +\infty} \sum_{j=1}^{k} \int_{D^j} W(\nabla u_n(x)(\nabla v_n)^{-1}(x)) \, dx$$

$$\leq \liminf_{n \to +\infty} \int_{\Omega} W(\nabla u_n(x)(\nabla v_n)^{-1}(x)) \, dx. \quad (6.35)$$

Letting $\eta$ go to zero, $k$ go to infinity, and then $\epsilon$ go to zero we obtain

$$E(u, v) \leq \liminf_{n \to +\infty} E(u_n, v_n).$$

*Case 2.* We assume that $1 < r = \frac{1}{p} + \frac{N-1}{q}$ and there are some constants $C_1, C_2 > 0, 1 \leq s \leq r$ such that $-C_1(1 + |F|_2^s) \leq W(F) \leq C_2(1 + |F|_2^r)$ for every $F \in M^{N \times N}$.

The proof follows as in the first case, where in (6.35) we use the lower semicontinuity results of Dacorogna (1987) instead of those of Fonseca and Müller (1992). Since $\{\nabla u_n(x)(\nabla v_n)^{-1}(x))\}$ is weakly relatively compact in $\Omega$, we have

$$\int_{\overset{k}{\underset{j=1}{\cup}} D^j(\eta)} W(\nabla u(x)(\nabla v)^{-1}(x)) \, dx = \sum_{j=1}^{k} \int_{D^j(\eta)} W(\nabla u(x)(\nabla v)^{-1}(x)) \, dx$$

$$= \sum_{j=1}^{k} \int_{B(y^j, r^j - \eta)} W((\nabla u \circ w^j)(y)) \, dy$$

$$\leq \sum_{j=1}^{k} \liminf_{n \to +\infty} \int_{B(y^j, r^j - \eta)} W((\nabla u_n \circ w_n^j)(y)) \, dy$$

$$= \sum_{j=1}^{k} \liminf_{n \to +\infty} \int_{D_n^j(\eta)} W(\nabla u_n(x)(\nabla v_n)^{-1}(x)) \, dx$$

$$= \sum_{j=1}^{k} \liminf_{n \to +\infty} [\int_{D^j} W(\nabla u_n(x)(\nabla v_n)^{-1}(x)) \, dx$$

$$+ \int_{D_n^j(\eta) \backslash D^j} W(\nabla u_n(x)(\nabla v_n)^{-1}(x)) \, dx$$

$$- \int_{D^j \backslash D_n^j(\eta)} W(\nabla u_n(x)(\nabla v_n)^{-1}(x)) \, dx]$$

$$\leq \sum_{j=1}^{k} \liminf_{n \to +\infty} \int_{D^j} W(\nabla u_n(x)(\nabla v_n)^{-1}(x)) \, dx$$

$$+ C_1 \int_{D^j \triangle D_n^j} (1 + |\nabla u_n(x)(\nabla v_n)^{-1}(x)|_2^s) \, dx$$

$$\leq \liminf_{n \to +\infty} \sum_{j=1}^{k} \int_{D^j} W(\nabla u_n(x)(\nabla v_n)^{-1}(x)) \, dx$$

$$\leq \liminf_{n \to +\infty} \int_{\Omega} W(\nabla u_n(x)(\nabla v_n)^{-1}(x)) \, dx.$$

Letting $\eta$ go to zero, $k$ go to infinity, and then $\epsilon$ go to zero we conclude that

$$E(u, v) \leq \liminf_{n \to +\infty} E(u_n, v_n).$$

$\square$

# 7

## DEGREE IN INFINITE-DIMENSIONAL SPACES

### 7.1 Introduction to the Leray–Schauder degree

We define the Leray–Schauder degree in a linear normed space of infinite dimension, $X$. Let $D \subset X$ be an open, bounded set, let $\phi : \bar{D} \to X$ be a continuous mapping, and let $p \in X \setminus \phi(\partial D)$. We want to define $d(\phi, D, p)$, expecting that

- (i) $d(I, D, p) = 1$ if $p \in D$, where $I$ stands for the identity mapping on $D$;
- (ii) $d(\phi, D, p) \neq 0$ implies that $p \in \phi(D)$;
- (iii) $d(h_t, D, p)$ is independent of $t$ if $h_t$ is a $C^0$ homotopy such that $p \notin h_t(\partial D)$, for every $t \in [0, 1]$.

In a finite-dimensional space $X$, $C^0(\bar{D}, X)$ is a suitable class of functions $\phi$ for which there is existence and uniqueness of a degree function satisfying (i)–(iii). In an infinite-dimensional space, an example given by Leray (1936) shows that one should restrict the class of functions to be strictly contained in $C^0(\bar{D}, X)$.

*Leray's example.* Let $X$ be the class of continuous functions $x : [0, 1] \to \mathbb{R}$ and for $x \in X$ we set

$$||x|| := \max\{|x(s)| \ : \ 0 \le s \le 1\}.$$

Consider

$$x_0(s) = \frac{1}{2}, \quad 0 \le s \le 1$$

and let $D \subset X$ be given by

$$D := \left\{ x \in X \ : \ ||x - x_0|| < \frac{1}{2} \right\}.$$

Then there exists $p \in X$ such that for any function $d(\cdot, D, p) : C^0(\bar{D}, X) \to \mathbb{Z}$ one of the properties (i)–(iii) fails. To see this, assume on the contrary that there exists $d(\cdot, D, p) : C(\bar{D}, X) \to \mathbb{Z}$ verifying (i), (ii) and (iii). Let $\gamma : [0, 1] \to [0, 1]$ be the continuous mapping defined by

$$\gamma(s) := \begin{cases} s & 0 \le s \le 1/2 \\ 1 - s & 1/2 \le s \le 5/8 \\ 5/3(s - 1) + 1 & 5/8 \le s \le 1 \end{cases}$$

and define $\phi : \bar{D} \to X$ by

$$\phi(x) := \gamma \circ x.$$

Since $\gamma([0, 1]) = [0, 1]$ we have $\phi(\bar{D}) \subset \bar{D}$, and we let $h_t(x) := tx + (1 - t)\phi(x)$, $0 \le t \le 1$, $x \in \bar{D}$. Clearly,

$$\|h_t(x) - x_0\| \le \frac{1}{2}.$$

*Claim 1.* $h_t(\partial D) \subset \partial D$ for every $t \in [0, 1]$.

Indeed, fix $t \in [0, 1]$ and if $x \in \partial D$, then

$$\|x - x_0\| = \frac{1}{2}, \tag{7.1}$$

which implies that

$$-\frac{1}{2} \le x(s) - x_0(s) \le \frac{1}{2} \text{ and } |x(s_0) - x_0(s_0)| = \frac{1}{2}$$

for every $s \in [0, 1]$ and some $s_0 \in [0, 1]$. We conclude that $x(s_0) \in \{0, 1\}$ and as $0 \le x(s) \le 1$ for every $s \in [0, 1]$, we have

$$|h_t(x)(s_0)| = \frac{1}{2},$$

which, together with the fact that $0 \le x(s) \le 1$ for every $s \in [1, 0]$, yields

$$0 \le h_t(x)(s) \le 1$$

for every $s \in [0, 1]$ and so

$$|h_t(x)(s) - x_0(s)| \le \frac{1}{2}.$$

Therefore,

$$\|h_t(x)\| = \frac{1}{2}.$$

Let $p \in X$ be defined by

$$p(s) := \frac{1}{4} + \frac{1}{2}.$$

Then

$$\|p - x_0\| = \frac{3}{8}$$

and so $p \notin \partial D$. By Claim 1 for every $t \in [0, 1]$, $p \notin h_t(\partial D)$ and so $d(h_t, D, p)$ is well defined. Using properties (i) and (iii),

$$d(\phi, D, p) = d(I, D, p) = 1$$

and by property (ii) we deduce that there exists $x \in D$ such that $p = \phi(x) = \gamma \circ x$.

*Claim 2.* The equation $x(s) = \frac{1}{2}$ admits exactly one solution.

Indeed,

$$\frac{1}{4} = p(0) \Rightarrow \frac{1}{4} = \gamma \circ x(0) \Rightarrow x(0) = \frac{1}{4}$$

and

$$\frac{3}{4} = p(1) \Rightarrow \frac{3}{4} = \gamma \circ x(1) \Rightarrow x(1) = \frac{17}{20}.$$

Therefore, by continuity the equation $x(s) = \frac{1}{2}$ admits at least one solution. Also,

$$x(s) = \frac{1}{2} \Rightarrow \gamma(x(s)) = \gamma\left(\frac{1}{2}\right) = \frac{1}{2} \Rightarrow p(s) = \frac{1}{2} \Rightarrow s = \frac{1}{2},$$

i.e. $s = \frac{1}{2}$ is the unique solution of the equation $x(s) = \frac{1}{2}$. Claim 2, together with the continuity of $x$, implies that either $x(t) > \frac{1}{2}$ for every $t \in (\frac{1}{2}, 1]$ or $x(t) < \frac{1}{2}$ for every $t \in (\frac{1}{2}, 1]$. Suppose that $x(s) < \frac{1}{2}$ for every $s \in (\frac{1}{2}, 1]$. Then for all $s \in (\frac{1}{2}, 1]$

$$x(s) < \frac{1}{2} \Rightarrow \gamma \circ x(s) < \frac{1}{2}$$

$$\Rightarrow p(s) < \frac{1}{2}$$

$$\Rightarrow \frac{1}{2} + \frac{1}{4} = p(1) < \frac{1}{2},$$

yielding a contradiction. We conclude that $x(s) > \frac{1}{2}$ for every $s \in (\frac{1}{2}, 1]$. Now, since $x(\frac{1}{2}) = \frac{1}{2}$ there exists $\epsilon > 0$ such that, for every $s \in [\frac{1}{2}, \frac{1}{2} + \epsilon]$, $x(s) \in [\frac{1}{2}, \frac{5}{8}]$. This implies $\gamma \circ x(s) \in [\gamma(\frac{5}{8}), \gamma(\frac{1}{2})] = [\frac{3}{8}, \frac{1}{2}]$. Recalling that $p(s) = \gamma \circ x(s) = \frac{1}{4} + \frac{s}{2}$ and that $p$ is increasing on $[\frac{1}{2}, \frac{1}{2}, +\epsilon]$, we obtain a contradiction.

Hence one of the properties (i), (ii), (iii) must fail for $d(\cdot, D, P)$.

In view of the above example, the Leray–Schauder degree is defined for compact perturbations of the identity, $I - T$, where $T$ is a compact mapping.

**Definition 7.1** Let $E, F$ be two normed spaces and let $M \subset E$. We say that $T : M \to F$ is compact if

(i) $T$ is continuous;

(ii) $\overline{T(A)}$ is a compact set for every $A \subset M$ bounded set.

**Definition 7.2** Let $E$ be a normed space and let $M \subset E$. We say that $M$ is of finite dimension if $M$ is contained in a linear subspace of $E$ of finite dimension.

**Theorem 7.3** Let $E, F$ be two normed spaces endowed, respectively, with the norms $||\cdot||_E$ and $||\cdot||_F$. Assume that $M \subset E$ is a bounded set and let $T : M \to F$ be a compact mapping. Then for every $\epsilon > 0$ there exists $T_\epsilon : M \to F$ such that $T_\epsilon(M)$ has finite dimension and $||T(x) - T_\epsilon(x)||_F < \epsilon$ for every $x \in M$.

**Proof** Since $M$ is bounded $\overline{T(M)}$ is a compact set. We have

$$\overline{T(M)} \subset \underset{p \in \overline{T(M)}}{\cup} B(p, \epsilon)$$

and, using the fact that $\overline{T(M)}$ is compact, we deduce that there exist points $p_1, \ldots, p_k \in \overline{T(M)}$ such that

$$\overline{T(M)} \subset \overset{k}{\underset{i=1}{\cup}} B(p_i, \epsilon).$$

Let

$$m_i(x, \epsilon) := \max\{0, \epsilon - ||T(x) - p_i||_F\}, \quad x \in M$$

and set

$$\theta_i(x, \epsilon) := \frac{m_i(x, \epsilon)}{\sum_{j=1}^{k} m_j(x, \epsilon)}, \quad x \in M.$$

We claim that $\theta_i(\cdot, \epsilon) : M \to \mathbb{R}$ is well defined and continuous, $i = 1, \ldots, k$. In fact, it is obvious that $m_i(\cdot, \epsilon)$ is continuous on $M$ and, given $x \in M$, there exists $j \in \{1, \ldots, k\}$ such that $T(x) \in B(p_j, \epsilon)$. Then

$$||T(x) - p_j||_F < \epsilon,$$

implying $m_j(x, \epsilon) > 0$; therefore

$$\sum_{l=1}^{k} m_j(x, \epsilon) > 0$$

and $\theta_i(\cdot, \epsilon)$ is well defined and continuous. Setting

$$T_\epsilon(x) := \sum_{j=1}^{k} \theta_j(x, \epsilon) p_j,$$

and since

$$\sum_{j=1}^{k} \theta_j(x, \epsilon) = 1,$$

we have

$$T(x) - T_\epsilon(x) = \sum_{j=1}^{k} \theta_j(x, \epsilon) T(x) - \sum_{j=1}^{k} \theta_j(x, \epsilon) p_j = \sum_{j=1}^{k} \theta_j(x, \epsilon)[T(x) - p_j].$$

Therefore,

$$||T(x) - T_\epsilon(x)||_F \leq \sum_{j=1}^{k} \theta_j(x, \epsilon) ||T(x) - p_j|| \leq \epsilon.$$

$\square$

**Lemma 7.4** *Let $X$ be a metric space, let $D \subset X$ be a bounded, open set, let $T : \bar{D} \to X$ be a compact mapping, and suppose that $p \notin \phi(\partial D)$, where $\phi := I - T$. Then $r := \rho(p, \phi(\partial D)) > 0$.*

**Proof** Let $\{x_k\} \subset \partial D$ be such that $r = \lim_{k \to +\infty} \rho(p, \phi(x_k))$. Since $\partial D$ is bounded and $T : \bar{D} \to X$ is compact, then $\{T(x_k)\}_k$ is relatively compact and so it admits a subsequence $\{T(x_{k_i}) : i \in \mathbb{N}\}$ such that $T(x_{k_i}) \to y \in X$. Assume that $r = 0$, i.e.

$$p = \lim_{k \to +\infty} \phi(x_k) = \lim_{k \to +\infty} (x_k - Tx_k).$$

Then

$$y + p = \lim_{r \to +\infty} x_{k_i} \in \partial D$$

and so

$$T(y + p) = \lim_{i \to +\infty} Tx_{k_i} = (y + p) - p = y.$$

Therefore,

$$\phi(y + p) = y + p - T(y + p) = y + p - y = p,$$

and we conclude that $p \in \phi(\partial D)$, which contradicts the hypothesis. $\qquad\square$

In the sequel, $X$ stands for a linear normed space, endowed with the norm $||\cdot||$ associated to the metric $\rho$, $I$ stands for the identity mapping on $X$, $T : \bar{D} \to X$ is a compact mapping, $\phi := I - T$, and $p \notin \phi(\partial D)$. Also, $D \subset X$ is a bounded, open set.

In view of Theorem 7.3, let $T_\epsilon : \bar{D} \to X$ be a compact mapping such that the smallest linear space span $T_\epsilon(\bar{D})$ containing $T_\epsilon(\bar{D})$ is of finite dimension and

$$\rho(T(x), T_\epsilon(x)) < \epsilon,$$

for every $x \in \bar{D}$. Set

$$S_\epsilon := \text{span}(T_\epsilon(\bar{D})), \quad D_\epsilon = D \cap S_\epsilon$$

and

$$\phi_\epsilon(x) := x - T_\epsilon(x).$$

**Lemma 7.5** *For every $0 < \epsilon < \rho(p, \phi(\partial D))$, $d(\phi_\epsilon, D_\epsilon, p)$ is well defined and is independent of $\epsilon$.*

**Proof** Recall that, by Lemma 7.4, $\rho(p, \phi(\partial D))r > 0$ for some $r > 0$. Let $\partial_\epsilon D_\epsilon$ be the boundary of $D_\epsilon$ in $S_\epsilon$. We have $\partial_\epsilon D_\epsilon \subset \partial D$ and $p \notin \phi_\epsilon(\partial_\epsilon D_\epsilon)$, for $0 < \epsilon < r$. Let $\epsilon_1, \epsilon_2 \in (0, r)$ and set

$$S_\nu := \text{span}(S_{\epsilon_1}, S_{\epsilon_2})$$

and

$$D_\nu := D \cap S_\nu.$$

By Lemma 1.22 we have

$$d(\phi_{\epsilon_1}, D_V, p) = d(\phi_{\epsilon_1}, D_{\epsilon_1}, p).$$

and

$$d(\phi_{\epsilon_2}, D_V, p) := d(\phi_{\epsilon_2}, D_{\epsilon_2}, p).$$

Clearly,

$$H(x, t) = t\phi_{\epsilon_1}(x) + (1 - t)\phi_{\epsilon_2}(x), \quad t \in [0, 1], x \in \bar{D}_V$$

is a $C^0$-homotopy between $\phi_{\epsilon_1}$ and $\phi_{\epsilon_2}$, and for every $t \in [0, 1]$, $p \notin H(\partial D, t)$. This yields $d(\phi_{\epsilon_2}, D_V, p) = d(\phi_{\epsilon_1}, D_V, p)$ and so

$$d(\phi_{\epsilon_1}, D_{\epsilon_1}, p) = d(\phi_{\epsilon_2}, D_{\epsilon_2}, p).$$

□

**Definition 7.6** *Let $T := \bar{D} \to X$ be a compact mapping, $\phi := I - T$, and let $p \notin \phi(\partial D)$. The Leray–Schauder degree of $\phi$ at $p$ with respect to $D$ is defined by $d(\hat{\phi}, D_V, p)$, where $\hat{\phi} := I - \hat{T}, \hat{T} : \bar{D} \to X$ is a compact mapping such that $\|\hat{T}(x) - T(x)\| < \rho(p, \phi(\partial D))$, $\hat{T}(\bar{D})$ is of finite dimension, $D_V = D \cap V$, and $V$ is any linear space of finite dimension containing $p$ and $\hat{T}(\bar{D})$.*

**Remark 7.7**

(i) Note that by Lemma 7.4, $p \notin (\phi(\partial D))$ implies that $\rho(p, \phi(\partial D)) > 0$.

(ii) It is easy to see that, in order to define $d(\phi, D, p)$, we only need to assume that $D \cap V$ is bounded in $V$ for every linear space $V \subset X$ of finite dimension, which allows us to extend the notion of degree to some unbounded spaces.

## 7.2   Properties of the Leray–Schauder degree

We prove that the Leray–Schauder degree satisfies most of the properties of the Brouwer degree. In the sequel $p, D, \phi$ are as in the Section 7.1 and $T := I - \phi$. If $M$ is a subset of $X$ we denote by $K(M)$ the set of compact mappings from $M$ into $X$ and

$$K_1(M) := \{\phi : \phi = I - T, T \in K(M)\}.$$

$K_1(M)$ is the set of *compact perturbations of the identity* on $M$.

**Theorem 7.8** *The following assertions hold.*

(i) $d(I, D, p) = 1$ *for every $p \in D$;*

(ii) $d(I, D, p) = 0$ *for every $p \notin \bar{D}$;*

(iii) $d(\phi, D, p) \neq 0$ *implies that there exists $x \in D$ such that $\phi(x) = p$.*

**Proof**

(i) Let $T_\epsilon(x) \equiv 0$, $x \in \bar{D}$, $\mathcal{S}_\epsilon := \text{span}\{p\}$ and set

$$D_\epsilon := D \cap \mathcal{S}_\epsilon.$$

We have

$$d(I, D, p) = d(I, D_\epsilon, p)$$

and as $p \in D$, then $p \in D_\epsilon$ and so, by the definition of Brouwer degree,

$$d(I, D_\epsilon, p) = 1.$$

(ii) Similarly, $d(I, D, p) = 0$ if $p \notin \bar{D}$.

(iii) For every $n > \frac{1}{\rho(p, \phi(\partial D))}$ there exists $T_n : \bar{D} \to X$ such that

$$||T_n(x) - T(x)|| < \frac{1}{n}$$

for every $x \in \bar{D}$, and $T_n \hat{D}$ is of finite dimension. Let

$$\mathcal{S}_n := \text{span}\{T_n(\bar{D}), p\}.$$

By Definition 7.6 we have

$$d(I - T, D, p) = d(I - T_n, D_n, p)$$

where $D_n := D \cap \mathcal{S}_n$. By hypothesis $d(I - T_n, D_n, p) \neq 0$ and so, by Theorem 2.1, there exists $x_n \in D_n$ such that

$$x_n - T_n(x_n) = p. \tag{7.2}$$

Since $T : \bar{D} \to X$ is a compact mapping and $x_n \in \bar{D}$, which is a bounded set, we may assume without loss of generality that the sequence $T(x_n)$ converges to some $y \in X$. Hence, by (7.2) $x_n$ converges to some $p + y \in \bar{D}$ and using the continuity of $\phi$ at $y + p$ we obtain

$$\phi(y + p) = \lim_{n \to +\infty} \phi(x_n)$$
$$= \lim_{n \to +\infty} x_n - T(x_n)$$
$$= p.$$

Since $p \notin \phi(\partial D)$, we have $y + p \in D$ and the equation $\phi(x) = p$ admits the solution $p + y$ in $D$.

$\square$

**Definition 7.9** *Let $M \subset X$ and $H : \bar{D} \times [0, 1] \to X$. We say that $H$ is a homotopy of compact transformations on $M$ if*

(i) $H(\cdot, t) \in K(M)$ for every $t \in [0, 1]$;

(ii) for every $\epsilon > 0$ and every bounded set bounded $L \subset M$ there exists $\delta > 0$ such that $\|H(x, t) - H(x, s)\| < \epsilon$ whenever $x \in L$ and $|s - t| < \delta$.

**Theorem 7.10 [Invariance Under Homotopy]** Assume that $H : \bar{D} \times [0, 1] \to X$ is a homotopy of compact transformations on $\bar{D}$. Set

$$\phi_t := I - H(\cdot, t)$$

for $t \in [0, 1]$ and assume that $p \notin \phi_t(\partial D)$, for every $t \in [0, 1]$. Then $d(\phi_t, D, p)$ is independent of $t$.

**Proof** We start by claiming that there exists $r > 0$ such that $\|p - \phi_t(x)\| \geq r$ for every $x \in \partial D$, $t \in [0, 1]$. Suppose, on the contrary, that there are $\{t_n\} \subset [0, 1]$ and $\{x_n\} \subset \partial D$ such that

$$\lim_{n \to +\infty} \|p - \phi_{t_n}(x_n)\| = 0.$$

Without loss of generality we may assume that the relatively compact sequence $\{t_n : n \in \mathbb{N}\}$ converges to a suitable $\tau \in [0, 1]$ and since $H(\cdot, \tau)$ is compact, let $\{H(x_n, \tau)\}$ converge to some $y \in X$. We have

$$\begin{aligned}
p &= \lim_{n \to +\infty} \phi_{t_n}(x_n) \\
&= \lim_{n \to +\infty} [x_n - H(x_n, \tau)] + [H(x_n, \tau) - H(x_n, t_n)] \\
&= \lim_{n \to +\infty} (x_n - y)
\end{aligned}$$

because, by Definition 7.9(ii), $\lim_{n \to +\infty} (H(x_n, \tau) - H(X_n, t_n)) = 0$. Hence,

$$p + y = \lim_{n \to +\infty} x_n \in \partial D$$

and by the continuity of $H(\cdot, \tau)$ we deduce that $p = (p + y) - H(p + y, \tau) = \phi_\tau(p + y)$, which yields a contradiction.

Let $\mathcal{R}$ be the relation defined on $[0, 1]$ by

$$t \mathcal{R} s \quad \text{if } d(\phi_t, D, p) = d(\phi_s, D, p).$$

It is clear that $\mathcal{R}$ is an equivalence relation.

Next we show that the equivalence classes of $\mathcal{R}$ are open sets of $[0, 1]$.

Let $s \in [0, 1]$ and let $C$ be the equivalence class of $s$ with respect to $\mathcal{R}$. Set $r := \rho(p, \phi_s(\partial D))$. By Lemma 7.4, $r > 0$ and we fix $\epsilon \in (0, \frac{r}{4})$. By Theorem 7.3 there exists a compact mapping $h_\epsilon : \bar{D} \to X$ such that $h_\epsilon(\bar{D})$ is of finite dimension and

$$\|h_\epsilon(x) - H(x, s)\| < \epsilon \tag{7.3}$$

for every $x \in \bar{D}$. Using Definition 7.9, there exists $\delta > 0$ such that $|t - s| < \delta$ implies that

$$\|H(x,t) - H(x,s)\| < \epsilon \tag{7.4}$$

for every $x \in \bar{D}$. Let $V_\epsilon$ be a space of finite dimension containing $p, h_\epsilon(\bar{D})$, and set

$$D_\epsilon := D \cap V_\epsilon.$$

By (7.3) and (7.4) we have

$$d(I - h_\epsilon, D_\epsilon, p) = d(I - H(\cdot, s), D, p) \tag{7.5}$$

and

$$\|h_\epsilon(x) - H(x,t)\| < 2\epsilon < r$$

for every $|t - s| < \delta$. Thus

$$d(I - h_\epsilon, D_\epsilon p) = d(I - H(\cdot, t), D, p) \tag{7.6}$$

which, together with (7.5), yields

$$d(I - H(\cdot, t), D, p) = d(I - H(\cdot, s), D, p).$$

Therefore, $t\mathcal{R}s$. We conclude that $C$ is an open set of $[0,1]$ and as $[0,1]$ is a connected set, $\mathcal{R}$ admits only one equivalence class; hence $0\mathcal{R}1$. $\qquad\square$

**Theorem 7.11** *Assume that* $\phi, \psi \in K_1(\bar{D}), \phi|_{\partial D} = \psi|_{\partial D}$, *and* $p \notin \phi(\partial D)$. *Then* $d(\phi, D, p) = d(\psi, D, p)$.

**Proof** Let $H(x,t) := tT(x) + (1 - t)S(x), t \in [0,1], x \in \bar{D}$, where $T := I - \phi$ and $S := I - \psi$. $H$ is a homotopy of compact transformations on $\bar{D}$ and

$$H(\cdot, t)|_{\partial D} = (I - \phi)|_{\partial D}$$

for every $t \in [0,1]$. Due to the invariance under homotopy (see Theorem 7.10) we have

$$d(\phi, D, p) = d(\psi, D, p).$$

$\qquad\square$

**Theorem 7.12** *Assume that* $\phi \in K_1(\bar{D}), p \notin \phi(\partial D)$, *and let* $q \in X$. *Then* $d(\phi, D, p) = d(\phi - q, D, p - q)$.

**Proof** Let $\hat{\phi} : \bar{D} \to X$ be such that $\hat{T} = I - \hat{\phi}$ is a compact mapping, $p, q, \hat{T}(\bar{D})$ are contained in a space of finite dimension $V$, and

$$\|\hat{T}(x) - T(x)\| < \rho(p, \phi(\partial D))$$

for every $x \in \bar{D}$. By Definition 7.6,

$$d(I - \hat{T}, D_V, p) = d(\phi, D, p), \tag{7.7}$$

where

$$D_V := D \cap V, \quad \text{span}\{\hat{T}(\bar{D}), p, p - q\} \subset V$$

and

$$\|(\hat{T} + q)(x) - (T + q)(x)\| < \rho(p - q, (\phi - q)(\partial D)).$$

Since span$\{\bar{T}(\bar{D}) + q, p - q\} \subset V$, we have

$$d(I - \hat{T} - q, D_V, p - q) = d(\phi - q, D, p - q) \tag{7.8}$$

and, by Proposition 2.5,

$$d(I - \hat{T} - q, D_V, p - q) = d(I - \hat{T}, D_V, p),$$

which, together with (7.7) and (7.8), yields

$$d(\phi, D, p) = d(\phi - q, D, p - q).$$

$\square$

**Theorem 7.13** Let $\phi, \psi \in K_1(\bar{D})$, $p \notin \phi(\partial D)$, and let $\|\phi(x) - \psi(x)\| < r := \rho(p, \phi(\partial D))$ for every $x \in \bar{D}$. Then $p \notin \psi(\partial D)$ and

$$d(\phi, D, p) = d(\psi, D, p).$$

**Proof** Let $H(x, t) := t\phi(x) + (1 - t)\psi(x)$, $x \in \bar{D}, 0 \le t \le 1$. $H$ is a homotopy of compact transformations and, for every $x \in \partial D$,

$$\begin{aligned} \|p - H(x, t)\| &\ge \|p - \phi(x)\| - (1 - t)\|\phi(x) - \psi(x)\| \\ &> \|p - \phi(x)\| - (1 - t)r \\ &> r - (1 - t)r = tr. \end{aligned}$$

Therefore, for every $t \in [0, 1]$ we have $p \notin H(\partial D, t)$ and so $p \notin \psi(\partial D)$. By Theorem 7.10 (invariance under homotopy) we deduce that

$$d(\phi, D, p) = d(\psi, D, p).$$

$\square$

**Theorem 7.14** Let $\phi \in K_1(\bar{D})$ and let $\Omega$ be a connected component of $X \setminus \phi(\partial D)$. Then $d(\phi, D, \cdot)$ is constant in $\Omega$.

**Proof** Define $f : \Omega \to \mathbb{Z}$ as $f(p) := d(\phi, D, p)$. It suffices to show that $f$ is continuous on $\Omega$.

Fix $p \in \Omega$ and set $r := \rho(p, \phi(\partial D))$. By Lemma 7.4 we have $r > 0$ and $B(p, r) \subset \Omega$. For $q \in \Omega$ define $\phi_q : \bar{D} \to X$ by

$$\phi_q(x) := \phi(x) - (q - p),$$

for every $x \in \bar{D}$. It is clear that $\phi_q \in K_1(\bar{D})$ and, by Theorem 7.12, we have

$$d(\phi, D, q) = d(\phi - (q - p), D, q - (q - p)) = d(\phi_q, D, p). \qquad (7.9)$$

If $|p - q| < r$, then $\|\phi(x) - \phi_q(x)\| < \rho(p, \phi(\partial D))$ and, by Theorem 7.13,

$$d(\phi_q, D, p) = d(\phi, D, p) \qquad (7.10)$$

which, together with (7.9) and (7.10), yields

$$d(\phi, D, p) = d(\phi, D, q).$$

Hence $f : \Omega \to \mathbb{Z}$ is continuous and since $\Omega$ is connected, we conclude that $f(\Omega) = \{f(p)\}$ for every $q \in \Omega$. $\qquad \square$

**Theorem 7.15** *Let $\phi \in K_1(\bar{D})$ and let $p \notin \phi(\partial D)$. Then*

(i) *(Domain decomposition property) If $D = \cup_{i \in \mathbb{N}} D_i$ and $D_i$ are open sets, mutually disjoint, then*

$$d(\phi, D, p) = \sum_{i=1}^{+\infty} d(\phi, D_i, p).$$

(ii) *(Excision property) If $K \subset \bar{D}$ is a compact set such that $p \notin \phi(K)$, then*

$$d(\phi, D, p) = d(\phi, D \setminus K, p).$$

**Proof** Approximate $T := I - \phi$ by $\hat{T} \in K_1(\bar{D})$ such that $\hat{T}(\bar{D})$ is of finite dimension and

$$\|\hat{T}(x) - T(x)\| < \rho(p, \phi(\partial D))$$

for every $x \in \bar{D}$. Now (i) and (ii) follow from the domain decomposition property and the excision property of the degree in finite-dimensional spaces (see Theorem 2.7). $\qquad \square$

**Proposition 7.16** *If $\phi \in K_1(\bar{D})$ then*

(i) *$\phi(E)$ is closed for every closed set, $E \subset \bar{D}$;*

(ii) *$\phi^{-1}(K)$ is compact for every compact set, $K \subset X$.*

**Proof** Set $T := I - \phi$.

(i) Assume that $\{x_n : n \in \mathbb{N}\} \subset E$ and $\{\phi(x_n)\}$ converges to some $p \in X$. Since $T$ is compact and $\bar{D}$ is a bounded set, there exists a subsequence $\{x_{n_i}\}$ such that $\{Tx_{n_i}\}$ converges to some $y \in X$. We have

$$p + y = \lim_{i \to +\infty} x_{n_i}.$$

Since $E$ is closed, $p + y \in E$ and $\phi(p + y) = p$, thus $p \in \phi(E)$.

(ii) Let $\{x_n : n \in \mathbb{N}\} \subset \phi^{-1}(K)$. There exists $\{x_{n_i}\} \subset \phi^{-1}(K)$ such that $\{Tx_{n_i}\}_i$ converges to some $y \in X$. We have $x_{n_i} - T(x_{n_i}) = \phi(x_{n_i}) \in K$ and since $K$ is a compact set, up to a subsequence (not relabelled), $x_{n_i} - Tx_{n_i}$ converges to some $p \in X$. Thus

$$\lim_{i \to +\infty} x_{n_i} = \lim_{i \to +\infty} x_{n_i} - T(x_{n_i}) + T(x_{n_i}) = p + y.$$

Since $\phi^{-1}(K)$ is a closed set, we have $p + y \in \phi^{-1}(K)$ and so $\phi^{-1}(K)$ is a compact set.

$\square$

**Proposition 7.17** *If $\phi \in K_1(\bar{D})$ is one-to-one, then $\phi^{-1} : \phi(\bar{D}) \to X$ is also a perturbation of the identity, i.e. $\phi^{-1} \in K_1(\phi(\bar{D}))$.*

**Proof** Set $T := I - \phi$. We have

$$(I + T \circ \phi^{-1}) \circ \phi = \phi + T = I$$

therefore, $\phi^{-1} = I + T \circ \phi^{-1}$.

First we show that $T \circ \phi^{-1}(E)$ is relatively compact for every $E \subset \phi(\bar{D})$ bounded. Indeed, as $\phi^{-1}(E) \subset \bar{D}$ is bounded and $T$ is compact, we have that $T \circ \phi^{-1}(E)$ is relatively compact.

To prove that $T \circ \phi^{-1}$ is continuous, we consider a sequence $\{y_n : n \in \mathbb{N}\} \subset \phi(\bar{D})$ converging to $y \in \phi(\bar{D})$. There exists a set $\{x_n : n \in \mathbb{N}\} \subset \bar{D}$ and a point $x \in \bar{D}$ such that

$$y_n = \phi(x_n) \quad \text{and} \quad y = \phi(x).$$

It suffices to show that

$$x_n = \phi^{-1}(y_n) \to x = \phi^{-1}(y).$$

In view of Proposition 7.16 and since $\{y_n : n \in \mathbb{N}\}$ is relatively compact, we observe that the set $\{\phi^{-1}(y_n) : n \in \mathbb{N}\}$ is also relatively compact. Hence we may extract a subsequence (not relabelled) such that $\{x_{n_i}\}$ converges to some $a \in \bar{D}$. Using the continuity of $\phi$ at $a$, we have

$$\phi(a) = \lim_{i \to \infty} \phi(x_{n_i}) = y = \phi(x)$$

and as $\phi$ is injective, we conclude that $a = x$.                    $\square$

**Proposition 7.18 [Invariance of Domain]** *Let $\phi \in K_1(\bar{D})$ be a one-to-one function. Then $\phi(D)$ is open.*

**Proof**  By Proposition 7.17, $\phi^{-1} \in K_1(\phi(\bar{D}))$ and so $\phi(D) = (\phi^{-1})^{-1}(D)$ is an open set.                                                                                    □

In the following theorem we will assume that $X$ is a *uniformly convex* Banach space, i.e. $X$ is a Banach space endowed with a strictly convex norm,

$$\|\lambda x + (1 - \lambda)y\| < \lambda\|x\| + (1 - \lambda)\|y\|$$

for every $x, y \in X$ linearly independent and $\lambda \in (0, 1)$. It is well known that Hilbert spaces are uniformly convex.

**Theorem 7.19 [Odd Mapping]** *Let $X$ be a Banach space and let $D \subset X$ be an open, bounded, symmetric set containing $0$. Assume that $\phi \in K_1(\bar{D})$ is an odd mapping and $0 \notin \phi(\partial D)$. Then $d(\phi, D, 0)$ is an odd number.*

**Proof**  Set $T := I - \phi$. By Theorem 7.3 there exists a linear space $V \subset X$ of finite dimension $\hat{T} \in K(\bar{D})$ such that $\hat{T}(\bar{D})$ is contained in $V$ and

$$\|\hat{T}(x) - T(x)\| < \rho(0, \phi(\partial D))$$

for every $x \in \bar{D}$. As $X$ is uniformly convex, we may define the projection $p : X \to V$ in a unique way by

$$\|p(x) - x\| := \inf\{\|y - x\| : y \in V\}.$$

Then

$$\|p(T(x)) - T(x)\| \leq \|\hat{T}(x) - T(x)\| < \rho(0, \phi(\partial D)),$$

$p \circ T$ is a compact mapping and $V$ contains $p \circ T(\bar{D})$. Thus, by Definition 7.6

$$d(\phi, D, 0) = d(I - p \circ T, \ D \cap V, 0),$$

where $I - p \circ T$ is odd. Since $V$ has finite dimension, the result now follows from the odd mapping theorem in finite dimensions (see Theorem 3.23 and Remark 3.24).                                                                                    □

**Theorem 7.20 [Multiplication]** *Let $X$ be a Banach space, let $D, M \subset X$ be two open, bounded sets, and let $\phi \in K_1(\bar{D})$ be such that $\phi(\bar{D}) \subset M$. Assume that $\psi \in K_1(\bar{M})$ is such that $p \notin \psi(\partial M) \cup \psi \circ \phi(\partial D)$. Then*

$$d(\psi \circ \phi, D, p) = \sum_{i \in I} d(\psi, \Delta_i, p) d(\phi, D, \Delta_i),$$

*where $\Delta_i, i \in I$, are the connected components of $M \setminus \phi(\partial D)$.*

**Proof**  The proof is similar to the proof of the Multiplication Theorem in finite-dimensional spaces (see Theorem 2.10) and it is left to the reader.                         □

**Remark 7.21** Since $\bar{D}$ is bounded and $\overline{(I - \phi)(\bar{D})}$ is a compact set, we deduce that $\phi(\bar{D})$ is bounded and there exists a set $M$ satisfying the assumptions of Theorem 7.20. Also, by Proposition 7.16, $\phi(\partial D)$ is closed and so $\Delta_i$ is open for every $i \in I$. Finally, if $X$ is not separable, then the family $\{\Delta_i : i \in I\}$ may not be countable. However, since $\psi^{-1}(p)$ is a compact set, by Proposition 7.16 and Theorem 7.8 (iii) $d(\psi, \Delta_i, p) \neq 0$ for only finitely many indices $i \in I$ and so $\sum_{i \in I} d(\psi, \Delta_i, p) d(\phi, D, \Delta_i)$ is meaningful.

**Theorem 7.22 [Separation Theorem]** *Let $X$ be a Banach space, $D_1, D_2 \subset X$ be two open, bounded sets and let $h \in K_1(\bar{D}_1)$ be a homeomorphism from $\bar{D}_1$ onto $\bar{D}_2$. Then either the complements $(\bar{D}_1)^c$ and $(\bar{D}_2)^c$ of $\bar{D}_1$ and $\bar{D}_2$ have the same finite number of connected components or both have infinitely many connected components.*

**Proof** $h^{-1} = I - S \in K_1(D_2)$. Thus the hypotheses are symmetric and the proof follows.                                                                    □

**Remark 7.23** In finite-dimensional spaces, if $K$ and $L$ are two bounded, closed sets we deduce that $K^c$ and $L^c$ have the same number of connected components. In infinite-dimensional spaces, we need to assume that $K$ and $L$ are the closure of open sets and so $\text{int}(K)$ and $\text{int}(L)$ are not empty.

**Theorem 7.24 [Homeomorphism]** *Let $X$ be a Banach space and let $D \subset X$ be an open, bounded set. Assume that $\phi \in K_1(\bar{D})$ is one to one and $p \in \phi(D)$. Then $d(\phi, D, p) = \pm 1$.*

**Proof** The proof is identical to that of Theorem 3.35 for finite dimensional spaces.                                                                    □

## 7.3   Fixed point theorems

We present some fixed points theorems formulated by Schauder in 1930. Throughout this section, $X$ is a Banach space.

**Theorem 7.25** *Let $S \subset X$ be a bounded, convex, closed set containing the origin in its interior and let $T \in K(S)$ be such that $T(S) \subset S$. Then $T$ has a fixed point.*

**Proof** Let $D = \text{int}(S)$. First we prove that

$$\text{if } x \in S, \, 0 \leq t < 1, \text{ then } tx \in D. \tag{7.11}$$

Since $0 \in \text{int}(S) = D$, there exists $\epsilon > 0$ such that

$$\bar{B}(0, \epsilon) \subset S.$$

We prove that $B(tx, (1-t)\epsilon) \subset S$. Indeed, let $y \in B(tx, (1-t)\epsilon)$. Then $y = tx + (1-t)\epsilon u$ for some $u \in X$ such that $\|u\| < 1$, and since $t \in [0,1]$, $\epsilon u \in S$, and $x \in S$, by the convexity of $S$ we have $y \in S$. Therefore,

$$B(tx, (1-t)\epsilon) \subset S$$

and so $tx \in D$. If $Tx = x$ for some $x \in \partial D$, then the proof is concluded. Assume that $T(x) \neq x$ for every $x \in \partial D$. Set

$$h_t(x) := x - tT(x), \quad x \in \bar{D}, \ 0 \leq t \leq 1.$$

$h_t$ is a homotopy of compact perturbations. Let us prove that $h_t(x) \neq 0$ for every $x \in \partial D$ and every $t \in [0,1]$. Indeed, if $h_t(x) = 0$ for some $x \in \partial D$ and some $t \in [0,1]$, then $x = tT(x)$ and since $T$ has no fixed point in $\partial D$, we deduce that $t \neq 1$ and, by (7.11), $x \in D$. This yields a contradiction. Therefore, $0 \notin h_t(\partial D)$ for every $t \in [0,1]$ and by Theorems 7.8 (i) and 7.10 we have

$$1 = d(I, D, 0) = d(I - T, D, 0).$$

Using Theorem 7.8 (iii), we deduce that there exists $x \in D$ such that

$$T(x) = x.$$

$\square$

The following is a generalization of Tietze's Extension Theorem for infinite-dimensional-range spaces.

**Theorem 7.26 [Dugundji]** *Let $X$ and $Y$ be two normed spaces, let $A \subset X$ be a closed set, and let $C \subset Y$ be a convex set. If $f : A \to C$ is continuous, then there exists a continuous extension $F : X \to C$ of $f$.*

**Proof** Denote by $\rho$ be the distance on $X$ and by $|| \cdot ||$ the norm on $Y$. We construct a partition of unity of $X \backslash A$ as follows.

For $x \in X \setminus A$, let $B_x$ be a ball of centre $x$ and radius $r(x) := \frac{\rho(x,A)}{6}$. We observe that

$$\text{diam}(B_x) \leq \rho(B_x, A), \ X \setminus A = \underset{x \in X \backslash A}{\cup} B_x.$$

Due to the paracompactness property of metrizable spaces, $\{B_x : x \in X \setminus A\}$ admits a locally finite refinement $\{U_j : j \in I\}$, i.e. there exists a family of open sets $\{U_j : j \in I\}$ which covers $X \setminus A$ such that for every $x \in X \setminus A$ there exists a neighbourhood $V(x)$ which meets only finitely many of the $U_j$ and for every $j \in I$ there exists $z \in X \setminus A$ such that $U_j \subset B_z$. Thus, given $j \in I$ and $B_x$ such that $U_j \subset B_x$,

$$\rho(U_j, A) \geq \rho(B_x, A) > 0$$

and so we can choose $a_j \in A$ such that

$$\rho(a_j, U_j) < 2\rho(U_j, A).$$

Define $\phi_i : X \setminus A \to \mathbb{R}$ by

$$\phi_i(x) := \begin{cases} \rho(x, \partial U_i) & \text{if} \quad x \in U_i \\ 0 & \text{if} \quad x \notin U_i \end{cases}$$

and $\psi_i : X \setminus A \to \mathbb{R}$ by

$$\psi_i(x) := \frac{\phi_i(x)}{\displaystyle\sum_{j \in I} \phi_j(x)}.$$

We observe that $\sum_{j \in I} \phi_j(x) > 0$ for every $x \in X \setminus A$. Since there are only finitely many $U_j$ which meet $V(x)$, we deduce that $\psi_i$ is continuous on $X \setminus A$. It is easy to see that

$$0 \leq \psi_i(x) \leq 1 \quad \text{and} \quad \sum_{j \in I} \psi_j(x) = 1$$

for every $x \in X \setminus A$. Set

$$F(x) := \begin{cases} f(x) & x \in A \\ \displaystyle\sum_{j \in I} \psi_j(x) f(a_j) & x \notin A. \end{cases}$$

Clearly, $F$ is an extension of $f$ and

$$F(X) \subset C.$$

We show that $F$ is continuous on $X$.

Indeed, since for every $x \in X \setminus A$ there exist only finitely many $U_j$ which meet $V(x)$ and $\psi_j$ is continuous for every $j \in I$, we conclude that $F$ is continuous at every point of $X \setminus A$. Finally, let $\epsilon > 0$ and $x_0 \in \partial A \subset A$. Fix $\delta > 0$ such that

$$\|f(x) - f(x_0)\| < \epsilon$$

for every $x \in A \cap B(x_0, \delta)$ and assume that $\rho(x, x_0) < \frac{\delta}{4}$. Either $\psi_j(x) = 0$ or $\psi_j(x) \neq 0$, in which case we have $x \in U_j \subset B_z$ for some $z \in X \setminus A$ and so

$$\begin{aligned} \rho(x, a_j) &\leq \rho(a_j, U_j) + \text{diam}(U_j) \\ &\leq 2\rho(A, U_j) + \text{diam}(B_z) \\ &\leq 3\rho(A, U_j) \\ &\leq 3\rho(x, x_0), \end{aligned}$$

implying

$$\rho(a_j, x_0) \leq 4\rho(x, x_0) < \delta.$$

Thus we have

$$\begin{aligned} \|F(x) - F(x_0)\| = \|\sum_{j \in I} \psi_j(x)(f(a_i) - f(x_0))\| \\ \leq \sum_{j \in I} \psi_j(x)\|f(a_i) - f(x_0)\| \end{aligned}$$

$$\leq \sum_{j\in I} \psi_j(x)\epsilon$$
$$= \epsilon$$

and $F$ is continuous at $x_0 \in \partial A$.                                    □

**Remark 7.27** In the case where $X = Y$ Theorem 7.26 was first proved by S. Katukani in 1951. The general case is obtained by Dugundji (1951), and in his original statement he only requires that $X$ be a metric space and $Y$ be a locally convex topological space.

**Lemma 7.28** *Let $X$ be a Banach space and let $K \subset X$ be a compact set. Then the convex hull of $K$ is a compact set.*

**Proof**  We divide the proof into two steps.
*Step 1.* We prove that for every $\epsilon > 0$ there exist finitely many balls of radius $\epsilon$ covering $C$.

Let $\epsilon > 0$. Since $K$ is a compact set, there exist $x_1, \ldots, x_n \in K$ such that

$$K \subset \overset{n}{\underset{i=1}{\cup}} B\left(x_i, \frac{\epsilon}{2}\right).$$

Let

$$D := \text{conv}\{x_1, \ldots, x_n\}, E = \left\{x \in X : \rho(x, D) \leq \frac{\epsilon}{2}\right\}.$$

We claim that $C \subset E$. Indeed, if $x \in C$, then there are $\lambda_1, \cdots, \lambda_k \in [0,1]$, $c_1, \ldots, c_k \in K$ such that

$$\sum_{i=1}^{k} \lambda_i = 1, \quad \sum_{i=1}^{k} \lambda_i c_i = x.$$

For each $j \in \{1, \ldots, k\}$, there exists $i(j)$ such that $||c_j - x_{i(j)}|| < \frac{\epsilon}{2}$. We have

$$x = \sum_{j=1}^{k} \lambda_j(c_j - x_{i(j)}) + \sum_{j=1}^{k} \lambda_j x_{i(j)},$$

which implies that

$$\left\|x - \sum_{j=1}^{k} \lambda_j x_{i(j)}\right\| < \frac{\epsilon}{2}$$

and so $x \in E$.

Since $D$ is a compact, convex set, there exist $y_i, \ldots, y_l \in D$ such that

$$D \subset \overset{l}{\underset{j=1}{\cup}} B\left(y_j, \frac{\epsilon}{8}\right). \tag{7.12}$$

We show that $C \subset \overset{l}{\underset{j=i}{\cup}} B(y_j, \epsilon)$.

Let $x \in C$. As $C \subset E$ there exists $a \in D$ such that $\|x - a\| \leq \frac{3}{4}\epsilon$ and so, by (7.12), we deduce that there exists $j \in \{1, \ldots, l\}$ such that

$$\|x - y_j\| \leq \frac{3}{4}\epsilon + \frac{\epsilon}{8} < \epsilon.$$

Hence,

$$C \subset \bigcup_{j=i}^{l} B(y_j, \epsilon).$$

*Step 2.* We conclude that $C$ is relatively compact. Let $\{x_n : n \in \mathbb{N}\} \subset C$. By Step 1, there exist $k_1$ balls $B_1^1, \ldots, B_{k_1}^1$ of radius $\frac{1}{2}$ covering $C$. One of them contains infinitely many $x_n$, say

$$x_n^1 \in B_1^1, \quad n \in \mathbb{N}.$$

By iteration, we extract from $\{x_n\}$ a subsequence in the following way : if $i \geq 2$, $B_1^i, \ldots, B_{k_2}^i$ is a family of open balls of radius $\frac{1}{2^i}$ covering $C$, $\{x_n^i\}$ is a subsequence of $\{x_n^{i-1}\}$ such that

$$x_n^i \in B_1^i, \quad n \in \mathbb{N}.$$

Consider the diagonal subsequence

$$y_n := x_n^n$$

For every $\epsilon > 0$ we have

$$\|y_n - y_m\| < \epsilon$$

whenever $n, m \geq 1 - \frac{\log \epsilon}{\log 2}$. Therefore $\{y_n\}$ is a Cauchy sequence and since $X$ is a Banach space, $\{y_n\}$ converges to some $x \in X$. This proves that $C$ is relatively compact. $\qquad\square$

**Corollary 7.29** *Let $X$ be a Banach space, let $D \subset X$ be an open, bounded set and let $T \in K(\bar{D})$. Then there exists a continuous extension $F$ of $T$ such that $F(X) \subset \operatorname{conv} T(\bar{D})$ and $F \in K(X)$.*

**Proof** Let $C := \operatorname{conv} K$ where $K = T(\bar{D})$. Since $K$ is a compact subset of a Banach space, by Lemma 7.28 we deduce that $C$ is a compact, convex set. By Theorem 7.26 there exists a continuous extension $F : X \to C$ of $T$ such that $F(X) \subset C$. Finally, since $C$ is a compact set, it follows that $F$ is a compact mapping. $\qquad\square$

Using the corollary of the Dugundjis Theorem, we next state and prove an improved version of Theorem 7.25, where we do not impose the condition $\operatorname{int}(S) \neq \emptyset$.

**Theorem 7.30** *Let $S \subset X$ be a nonempty, bounded, closed, convex set and let $T \in K(S)$ verify $T(S) \subset S$. Then $T$ admits a fixed point.*

**Proof** Since $S$ is bounded there exists $r > 0$ such that $S \subset B(0, r)$. By Corollary 7.29 there exists a compact extension $F : \bar{B}(0, r) \to S \subset \bar{B}(0, r)$ of $T$ and by Corollary 7.25 there exists $x \in \bar{B}(0, r)$ such that

$$F(x) = x.$$

Since $F(x) \in S$, we have $x \in S$ and $F(x) = T(x)$. Hence $T$ admits a fixed point.

$\square$

**Remark 7.31**

(i) Due to Theorem 7.25 every compact, convex, nonempty set $S \subset X$ has the fixed point property, since every continuous mapping $T : S \to S$ is compact.

(ii) Dugundji proved that the unit ball of $X$ has the fixed point property if and only if $X$ is finite dimensional. Therefore, Brouwer's Fixed Point Theorem (see Corollary 3.8) is inappropriate to infinite-dimensional spaces, because the unit ball of a normed $X$ of infinite dimension is not compact (see Exercise 7.1).

## 7.4    An application of the degree theory to ODEs

Throughout this section $a$, $c$ are two real numbers, $a < c$, and

$$X := \{x : [a, c] \to \mathbb{R} : x \text{ continuous}\}$$

is endowed with the norm

$$||x|| = \max\{|x(t)| : t \in [a, c]\}.$$

We consider a continuous function

$$g : [a, c] \times \mathbb{R} \to \mathbb{R}$$

and, given $k > 0$, we set

$$b := a + \min\left\{c - a, \frac{k}{M}\right\},$$

where

$$M := \max\{|g(s, u)| : a \le s \le c, \ |u| \le k\}.$$

**Lemma 7.32** *Let $D = B(0, k)$ be the open ball of centre at the origin and radius $k$ in $X$ and let $T : \bar{D} \to X$ be defined by*

$$T(x)(t) := \int_a^t g(s, x(s)) \, ds, \quad a \le t \le b.$$

*Then $T$ is a compact mapping.*

**Proof** First we prove that $T$ is continuous.

Let $\epsilon > 0$. Since $g|_{[a,b] \times [-k,k]}$ is uniformly continuous, there exists $\delta > 0$ such that

$$|g(s, u) - g(s, v)| < \frac{\epsilon}{b - a}$$

for every $s \in [a, b]$ and for every $|u - v| < \delta$. If $\|x - y\| < \delta$, then

$$|T(x)(t) - T(y)(t)| = \left| \int_a^t [g(s, x(s)) - g(s, y(s))] \, ds \right|$$
$$\leq \epsilon$$

for every $t \in [a, b]$ and so

$$\|T(x) - T(y)\| < \epsilon.$$

We prove that $T(\bar{D})$ is relatively compact. Clearly,

$$T(\bar{D}) \subset \bar{B}(0, (b - a)M). \tag{7.13}$$

On the other hand, if $x \in \bar{D}$, $t_1, t_2 \in [a, b]$, then

$$|T(x)(t_1) - T(x)(t_2)| = \left| \int_{t_2}^{t_1} g(s, x(s)) \, ds \right| \leq M|t_1 - t_2|$$

and so $T(\bar{D})$ is equicontinuous. This, together with the fact that $T(\bar{D})$ is bounded and by the Ascoli–Arzela Theorem, yields that $T(\bar{D})$ is relatively compact. Thus $T$ is compact. $\qquad \square$

**Theorem 7.33** *The first-order ordinary differential equation*

$$\begin{cases} \dot{x} & = g(t, x), \quad a < t < b \\ x(a) = 0 \end{cases}$$

*admits a solution* $x \in C^1([a, b])$.

**Proof** Let $x \in \bar{D} = \bar{B}(0, k)$. For every $t \in [a, b]$ we have

$$|T(x)(t)| = \left| \int_a^t g(s, x(s)) \, ds \right| \leq (b - a)M \leq \frac{k}{M} Mk$$

and so $T(\bar{D}) \subset \bar{D}$. As $T$ is a compact mapping (see Lemma 7.32) by the fixed point theorem (Theorem 7.30) there exists $x \in \bar{D}$ such that $T(x) = x$, i.e.

$$x(t) = \int_a^t g(s, x(s)) \, ds, \quad t \in [a, b].$$

Since $s \mapsto g(s, x(s))$ is continuous on $[a, b]$, $t \mapsto \int_a^t g(s, x(s)) \, ds = x(t) \in C^1[a, b]$ and we have

$$\begin{cases} \dot{x}(t) = g(t, x(t)), & a \le t \le b \\ x(a) = 0. \end{cases}$$

□

**Remark 7.34** If we assume only that $g : [a, c] \times \mathbb{R} \to \mathbb{R}$ is continuous, it is well known that, in general, the solution of the above ODE is not unique.

## 7.5    First application of the degree theory to PDEs

Throughout this section, $B \subset \mathbb{R}^N$ is the unit ball centred at 0, $B = B(0, 1)$, $\varphi \in C^1(\bar{B} \times \mathbb{R})$ is a bounded function, and $\Delta$ stands for the Laplacian operator. We want to solve the P.D.E

$$\begin{cases} \Delta x(u) = \varphi(u, x(u)) & u \in B \\ x(u) = 0 & u \in \partial B. \end{cases}$$

Let $Y = C^0(\bar{B})$ be the metric space endowed with the norm $||y||_Y := \max\{|y(u)| : u \in Y\}$, let $X = C^1(\bar{B})$ be the metric space with the norm $||x||_X := ||x||_Y + \max\{||\frac{\partial x}{\partial u_j}||_Y : j = 1, \ldots, N\}$ and let $Z := C^2(\bar{B})$ be the metric space with the norm

$$||x||_Z := ||x||_Y + \max\left\{\left|\left|\frac{\partial x}{\partial u_j}\right|\right|_Y : j = 1, \ldots, N\right\}$$

$$+ \max\left\{\left|\left|\frac{\partial^2 x}{\partial u_i \partial u_j}\right|\right|_Y : i, j = 1, \ldots, N\right\}.$$

It is well known that for $k = 0, 1, 2$, $C^k(\bar{B}, \mathbb{R})$ is a Banach space.

**Lemma 7.35** The inclusions $J_k : C^{k+1}(\bar{B}) \to C^k(\bar{B})$, defined by $J_k(x) = x$ for $k = 0, 1$, are compact mappings.

**Proof** We start by showing that $J_0$ is a compact mapping.
Since

$$||J(x)||_Y = ||x||_Y \le ||x||_X$$

for every $x \in X$, we conclude that $J_0(F)$ is bounded in $Y$ for every bounded set of $X, F \subset X$. On the other hand, $J_0(F)$ is equicontinuous for every bounded set $F \subset \bar{B}(0, M) \subset X$.

Indeed, let $x \in F$ and $u, v \in \bar{B}$. We have

$$x(u) - x(v) = \sum_{i=1}^{N} \frac{\partial x}{\partial u_i}(a)(u_i - v_i)$$

for some $a \in \bar{B}$ and so

$$|x(u) - x(v)| \le MN|u - v|.$$

This implies that $J_0(F)$ is equicontinuous and so, by the Ascoli–Arzela Theorem, $J_0(F)$ is relatively compact in $Y$. Thus $J_0$ is a compact mapping.

It can be proved in a similar way that $J_1$ is also a compact mapping.    □

**Lemma 7.36** *Let $A : X \to X$ be defined by*

$$A(x)(u) = \varphi(u, x(u)), \quad u \in \bar{B}.$$

*Then $A$ is continuous and the image by $A$ of any bounded set is a bounded set.*

**Proof**    First we remark that for every $x \in X$ and for every $u \in \bar{B}$

$$\frac{\partial A(x)}{\partial u_i}(u) = \frac{\partial \varphi}{\partial u_i}(u, x(u)) + \frac{\partial \varphi}{\partial x}(u, x(u)) \frac{\partial x}{\partial u_i}.$$

*Step 1.* We prove that $A(F)$ is bounded for every bounded set $F \subset X$.

Let $F \subset X$ and $M > 0$ be such that $\|x\|_X \leq M$ for every $x \in F$. We have

$$\|A(x)\|_X \leq L_1 + L_2 + L_3,$$

where

$$L_1 := \max \left\{ \left| \frac{\partial \varphi}{\partial u_i}(v, y) \right| : (v, y) \in E, \ i = 1, \dots, N \right\},$$

$$L_2 := \max \left\{ \left| \frac{\partial \varphi}{\partial u_i}(v, y) \right| : (v, y) \in E \ i = 1, \dots, N \right\},$$

$$L_3 := \max\{|\varphi(v, y)| : (u, y) \in E\},$$

$$E := \bar{B} \times [-M, M].$$

Thus $A(F)$ is a bounded set.

*Step 2.* We prove that $A$ is continuous on $X$.

Let $\{x_n : n \in \mathbb{N}\} \subset X$ be a sequence converging to some $x \in X$. Then there exists a constant $M > 0$ such that $\sup_n \|x_n\| \leq M$. Let $E, L_1, L_2,$ and $L_3$ be as in Step 1. Since $E$ is a compact set, $\varphi, \frac{\partial \varphi}{\partial u_i},$ and $\frac{\partial \varphi}{\partial x}$ are uniformly continuous on $E$ and so, fixing $\epsilon > 0$, there exists $\delta_1 \equiv \delta(\epsilon)$ such that for every $(u, y_1), (u, y_2) \in E$, such that $|y_1 - y_2| < \delta_1$, we have

$$|\varphi(u, y_1) - \varphi(u, y_2)| < \frac{\epsilon}{3}, \tag{7.14}$$

$$\max_{1 \leq i \leq N} \left| \frac{\partial \varphi}{\partial u_i}(u, y_1) - \frac{\partial \varphi}{\partial u_i}(u, y_2) \right| < \frac{\epsilon}{3}, \tag{7.15}$$

$$\left| \frac{\partial \varphi}{\partial x}(u, y_1) - \frac{\partial \varphi}{\partial x}(u, y_2) \right| < \frac{\epsilon}{6M}. \tag{7.16}$$

Let $\delta := \min \left\{ \delta_1, \frac{\epsilon}{6L_2} \right\}$ and $n_0 \equiv n_0(\epsilon)$ such that $n \geq n_0$ implies that

$$\|x_n - x\| \leq \delta.$$

For every $n \geq n_0$, by (7.14) we have

$$|A(x_n)(u) - A(x)(u)| \leq \frac{\epsilon}{3}.$$

By (7.15) and (7.16) we obtain

$$\left| \frac{\partial A(x_n)(u)}{\partial u_i} - \frac{\partial A(x)(u)}{\partial u_i} \right| \leq \left| \frac{\partial \varphi}{\partial u_i}(u, x_n(u)) - \frac{\partial \varphi}{\partial u_i}(u, x(u)) \right|$$
$$+ \left| \frac{\partial \varphi}{\partial x}(u, x_n(u)) \frac{\partial x_n}{\partial u_i} - \frac{\partial \varphi}{\partial u_i}(u, x(u)) \frac{\partial x}{\partial u_i} \right|$$
$$\leq \frac{2\epsilon}{3}.$$

Therefore, $\|A(x_n) - A(x)\|_X \leq \epsilon$ for every $n \geq n_0$ and we conclude that $A$ is continuous in $X$. $\qquad\square$

Let $u = (u_1, \ldots, u_N), v = (v_1, \ldots, v_N) \in \mathbb{R}^N$. As before,

$$|u|_2^2 = u_1^2 + \ldots + u_N^2 \quad \text{and} \quad u \cdot v = u_1 v_1 + \ldots + u_N v_N$$

and we set

$$|u, v| := [|u|_2^2 |v|_2^2 - 2u \cdot v + 1]^{\frac{1}{2}} = \begin{cases} \left| \frac{u}{|u|_2} - |u|_2 v \right| & u \neq 0 \\ 1 & u = 0. \end{cases}$$

Since $(|u|_2^2 - 1)(|v|_2^2 - 1) \geq 0$ for every $u, v \in \bar{B}$, we have

$$|u, v| \geq |u - v|_2 \tag{7.17}$$

with equality if and only if $u \in \partial B$ or $v \in \partial B$.

If $N = 2$ we define

$$G_1(u, v) := \frac{1}{\pi} \log|u - v|_2, \quad G_2(u, v) = \frac{1}{\pi} \log|u, v|, \quad u \neq v,$$

and if $N \geq 3$ let

$$G_1(u, v) := \frac{-1}{2(N-2)w_N |u - v|_2^{N-2}}, \quad u \neq v,$$

$$G_2(u, v) := \frac{-1}{2(N-2)w_N |u, v|_2^{N-2}}, \quad u \neq v,$$

where $w_N = \mathcal{L}^N(B)$. Define

$$G := G_1 - G_2.$$

For $a \in B$ and $r > 0$, we set $E(a, r) = B \setminus B(a, r) \cap B$ and $S(a, r) = \partial B(a, r)$ and we define

$$U := \{(u, v) : u, v \in \mathbb{R}^N, \ u \neq v\}.$$

**Lemma 7.37** *Let $F \in C^1(U)$ and let $L > 0$ be such that*

$$|F(u,v)|_2 \leq \frac{L}{|u-v|_2^{N-2}}, \quad |\nabla_v F(u,v)|_2 \leq \frac{L}{|u-v|_2^{N-1}} \qquad (7.18)$$

*for every $(u,v) \in U$. Then $v \mapsto \int_B F(u,v)\,du$ is continuously differentiable on $B$ and*

$$\frac{\partial}{\partial v_i} \int_B F(u,v)\,du = \int_B \frac{\partial F}{\partial v_i}(u,v)\,du.$$

**Proof**  Using Green's formula, we observe that

$$\frac{\partial}{\partial v_i} \int_{E(v,r)} F(u,v)\,du = \int_{E(v,r)} \frac{\partial F}{\partial v_i}(u,v)\,du - \int_{S(v,r)} F(s,v)n_i(s)\,ds,$$

where $n(s)$ denotes the outward unit normal to $S(v,r)$ at $s$. It is easy to see that, due to (7.18), by the continuity of $F$ and $\nabla_v F$ and by the Lebesgue Dominated Converging Theorem $\int_{E(v,r)} F(u,v)\,du$ converges uniformly in $v \in \bar{B}$ to $\int_B F(u,v)\,du$ when $r$ tends to zero, $\int_{E(v,r)} \frac{\partial F}{\partial v_i}(u,v)\,du$ converges uniformly in $v \in \bar{B}$ to $\int_B \frac{\partial F}{\partial v_i}(u,v)\,du$ when $r$ tends to zero, and $\int_{S(v,r)} F(s,v)n_i(s)\,ds$ converges uniformly in $v \in \bar{B}$ to $0$ when $r$ tends to zero. Therefore, letting $r \to 0^+$ we conclude that

$$\frac{\partial}{\partial v_i} \int_B F(u,v)\,du = \int_B \frac{\partial F}{\partial v_i}(u,v)\,du.$$

$\square$

**Lemma 7.38** *Let $F : C^1(\bar{B}) \to C^2(\bar{B})$ be defined by*

$$F(f)(v) := \int_B G(u,v)f(u)\,du.$$

*Then*

(i)

$$\begin{cases} \Delta x(u) = f(u) & u \in B \\ x(u) = 0 & u \in \partial B, \end{cases}$$

*where we have set $x = F(f)$.*

(ii) *There exists a constant $C > 0$ such that*

$$\|F(f)\|_Z \leq C\|f\|_X$$

*and*

$$\|F(f)\|_X \leq C\|f\|_Y,$$

*where $X = C^1(\bar{B})$, $Y = C^0(\bar{B})$, $Z = C^2(\bar{B})$.*

**Proof** We divide the proof into four main steps.

*Step 1.* We estimate $G_1$, $G_2$ and their partial derivatives.

There exists a constant $L > 0$ such that (see Exercise 7.2)

$$|G_1(u,v)|_2 \leq \frac{L}{|u-v|_2^{N-2}}, \text{ and } |G_2(u,v)|_2 \leq \frac{L}{|u,v|^{N-2}} \qquad (7.19)$$

for every $u, v \in \bar{B}$ such that $u \neq v$. Setting

$$\nabla_u G_i := \left( \frac{\partial G_i}{\partial u_j}, \ldots, \frac{\partial G_i}{\partial u_N} \right), i = 1, 2, \; j = 1, \ldots, N,$$

we obtain

$$|\nabla_u G_1(u,v)|_2 \leq \frac{L}{|u-v|_2^{N-1}} \text{ and } |\nabla_u G_2(u,v)|_2 \leq \frac{L}{|u,v|^{N-1}} \qquad (7.20)$$

for every $u, v \in \bar{B}$ such that $u \neq v$. Also, if

$$\nabla_u^2 G_i := \left( \frac{\partial G_i}{\partial u_j \partial u_k} \right), \; i = 1, 2 \; j, k = 1, \ldots, N,$$

then

$$|\nabla_u^2 G_1(u,v)|_2 \leq \frac{L}{|u-v|_2^N} \text{ and } |\nabla_u^2 G_2(u,v)|_2 \leq \frac{L}{|u,v|^N} \qquad (7.21)$$

for every $u, v \in \bar{B}$ such that $u \neq v$. In addition (see Exercise 7.3), for every $u, v \in \bar{B}$ such that $u \neq v$, we have

$$\Delta_u G_i(u,v) = 0 \; i = 1, 2 \qquad (7.22)$$

where

$$\Delta_u G_i = \sum_{j=1}^{N} \frac{\partial^2 G_i}{\partial u_j^2}.$$

Set

$$x(v) := \int_B G(u,v) f(u) \, du.$$

By (7.19) and (7.17), $x(v)$ is well defined.

*Step 2.* We prove that $x(v) = 0$ for every $v \in \partial B$.

In fact, if $v \in \partial B$ and if $u \in \bar{B}$, we have

$$|u, v| = |u - u|_2$$

and so

$$G(u,v) = G_1(u,v) - G_2(u,v) = 0,$$

which implies that

$$x(v) = 0.$$

*Step 3.* We prove that $\Delta x(v) = f(v)$ for every $v \in B$.
   Applying Lemma 7.37 to $G(u,v)f(u)$, we obtain

$$\frac{x(v)}{\partial v_i} = \int_B \frac{\partial G(u,v)}{\partial v_i} f(u)\, du =: y_i(v).$$

It is easy to see that $y_i$ is continuous on $B$ and

$$y_i(v) = \int_B \frac{\partial G(u,v)}{\partial v_i}[f(u) - f(v)]\, du + \int_B \frac{G(u,v)}{\partial v_i} f(v)\, du$$

$$= \int_B \frac{\partial G(u,v)}{\partial v_i}[f(u) - f(v)]\, du + f(v)\frac{v_i}{N}.$$

Since $f \in C^1(\bar{B})$ there exists a constant $K > 0$ such that

$$|f(u) - f(v)|_2 \le K|u - v|_2$$

for every $u, v \in \bar{B}$. Thus we can still apply Lemma 7.37 to $\frac{\partial G(u,v)}{\partial v_i}[f(u) - f(v)]$
to obtain

$$\frac{\partial}{\partial v_j} \int_B \frac{\partial G(u,v)}{\partial v_i}[f(u) - f(v)]\, du$$

$$= \int_B \frac{\partial^2 G(u,v)}{\partial v_i \partial v_j}[f(u) - f(v)]\, du - \int_B \frac{\partial G(u,v)}{\partial v_i}\frac{\partial f(v)}{\partial v_j}\, du$$

$$= \int_B \frac{\partial^2 G(u,v)}{\partial v_i \partial v_j}[f(u) - f(v)]\, du - \frac{\partial f(v)}{\partial v_j}\frac{v_i}{N}.$$

We deduce that

$$\frac{\partial^2 x(v)}{\partial v_i \partial v_j} = \int_B \frac{\partial^2 G(u,v)}{\partial v_i \partial v_j}[f(u) - f(v)]\, du + \frac{\delta_{ij}}{N} f(v), \qquad (7.23)$$

hence $\frac{\partial^2 x(u,v)}{\partial v_i \partial v_j}$ is continuous and (7.22) together with (7.23) yield

$$\Delta x(v) = f(v)$$

for every $v \in B$.
*Step 4.* We prove that $\|x\|_Z \le C\|f\|_X$ and $\|x\|_X \le C\|f\|_Y$.
   By (7.19), (7.20) and (7.21) we have

$$C_1 := \sup\left\{ \int_D |G(u,v)|\, du \ : \ u \in \bar{B} \right\} < +\infty,$$

$$C_2 := \sup \left\{ \int_D \left| \frac{\partial G(u,v)}{\partial v_i} \right| du \; : \; u \in \bar{B}, \; i = 1, \dots, N \right\} < \infty$$

and

$$C_3 := \sup \left\{ \int_B \left| \frac{\partial^2 G(u,v)}{\partial v_i \partial v_j} \right| \frac{1}{|u-v|_2} du \; : \; i, j = 1, \dots, N \right\} < \infty.$$

Therefore,

$$\|x\|_X \leq (C_1 + C_2)\|f\|_Y$$

and

$$\|x\|_Z \leq (C_1 + C_2)\|f\|_Y + C_3\|f\|_X.$$

Setting $C := C_1 + C_2 + C_3$, we have

$$\|x\|_X \leq C\|f\|_Y \quad \text{and} \quad \|x\|_Z \leq C\|f\|_X.$$

$\square$

**Lemma 7.39** *The equation*

$$\begin{cases} \Delta x(u) = \varphi(u, x(u)) & u \in B \\ x(u) = 0 & u \in \partial B \end{cases}$$

*admits a solution $u \in C^2(\bar{B})$.*

**Proof** Recall that by Lemma 7.35 the imbedding $J_1 : Z = C^2(\bar{B}) \to X = C^1(\bar{B})$ is a compact mapping, and let $T : X \to X$ be defined by

$$T := J_1 \circ F \circ A,$$

where $A(x)(u) := \varphi(x, u(x))$. We start by showing that $T$ is a compact mapping. Indeed, $T$ is continuous on $X$ because $J_1, A, F$ are continuous mappings. Moreover, if $S \subset X$ is a bounded set, then by Lemma 7.36 $A(S)$ is a bounded set of $X$ and so, using Lemma 7.38, $F \circ A(S)$ is a bounded set of $C^2(\bar{B})$. Since $J_1$ is a compact mapping, we obtain that $J_1 \circ F \circ A(S)$ is relatively compact in $X$, thus $T$ is a compact mapping.

Next we prove that $T(\bar{D}) \subset \bar{D}$, where $\bar{D} \subset X$ is the closed ball of radius $CM$, centred at 0, $C$ is the constant of Lemma 7.38, and

$$M := \sup\{|\varphi(u, z)| : u \in \bar{B}, z \in \mathbb{R}\}.$$

For $x \in \bar{D}$ we have

$$\|J_1 \circ F \circ A(x)\|_X = \|F \circ A(x)\|_X \leq C\|A(x)\|_Y \leq CM$$

and this proves that

$$T(\bar{D}) \subset (\bar{D}).$$

By the Leray–Schauder Fixed Point Theorem (see Theorem 7.25), there exists $x \in C^1(\bar{B})$ such that

$$Tx = x,$$

i.e.

$$F \circ A(x) = x.$$

As $F(y) \in C^2(\bar{B})$ for every $y \in C^1(\bar{B})$, we conclude that $x \in C^2(\bar{B})$ and, by Lemma 7.38,

$$\Delta x(u) = \Delta(F \circ A(x))(u) = Ax(u) = \varphi(u, x(u))$$

for every $u \in B$ and

$$x(u) = F \circ A(x)(u) = 0$$

for every $u \in \partial B.$                                                      □

## 7.6   Second application of the degree theory to PDEs

Consider the Dirichlet boundary value problem

$$\begin{cases} -\Delta u = \lambda f(x, u) & x \in D \\ u \quad\;\; = 0 & x \in \partial D, \end{cases} \tag{7.24}$$

where $D \subset \mathbb{R}^N$ is a bounded domain with smooth boundary, $\lambda > 0$, and $f \in C(\bar{D} \times [0, +\infty))$. If $f(x, 0) \geq 0$, then (7.24) is said to be a *positone problem*, and there is an extensive literature on this subject (see Amann 1976, Ambrosetti and Hess 1980, and Hess and Kato 1980). We say that (7.24) is *semi-positone* if

$$f(x, 0) < 0 \;\text{ for all } x \in D. \tag{7.25}$$

Work on this class of problems can be found in Castro, Garner, and Shivaji (1993), and Smoller and Wasserman (1987), among others. For related applications of the degree theory to boundary value problems, we refer to Rabinowitz (1971, 1975).

   We will show that a rescaling argument and the theory of Leray–Schauder topological degree will enable us to obtain positive solutions for the semi-positone problem (7.24) for certain values of $\lambda$, provided $f(x, \cdot)$ grows linearly at infinity. Precisely, we assume that there exists $m > 0$ such that

$$\lim_{u \to +\infty} \frac{f(u)}{u} = m, \tag{7.26}$$

and we define

$$\lambda_\infty := \frac{\lambda_1}{m},$$

where $\lambda_1$ is the first eigenvalue of $-\Delta$ with zero boundary conditions. Let $\varphi_1$ be the corresponding eigenfunction, with $\varphi_1 > 0$ in $D$ and $\|\varphi\|_{L^2(D)} = 1$. As in the previous section, we denote by $F$ the *Green operator* for $-\Delta$ with zero Dirichlet boundary conditions; $F : C(\bar{D}) \to C(\bar{D})$ is a continuous, compact operator such that $v = F(g)$ if and only if

$$\begin{cases} -\Delta v = g(v) & x \in D \\ v = 0 & x \in \partial D \end{cases}.$$

In the case where $f = f(u)$, we will show that $\lambda_\infty$ is a *bifurcation from infinity*, i.e. there exist $\lambda_n \to \lambda_\infty$ and $u_n \in C(\bar{D})$ such that $||u||_n \to +\infty$ and $u_n - \lambda_n F(f(|u_n|)) = 0$. Setting $w_n := u_n ||u_n||^{-1}$, elliptic theory and (7.26) yield that, up to a subsequence,

$$w_n \to w \quad \text{in } C^{1,\alpha}(\bar{D})$$

for some $0 < \alpha < 1$, where $||w|| = 1$ and

$$\begin{cases} -\Delta w = \lambda_1 |w| & x \in D \\ w = 0 & x \in \partial D \end{cases}. \tag{7.27}$$

By the maximum principle $w \geq 0$ and as $w \neq 0$, we must have $w = \gamma \varphi_1$, for some $\gamma > 0$. Thus $u_n > 0$ in $D$ for $n$ large.

Ambrosetti, Arcoya, and Buffoni (1994) used this argument to prove the theorem below, where

$$a(x) := \liminf_{u \to +\infty} (f(x, u) - mu), \quad A(x) := \limsup_{u \to +\infty} (f(x, u) - mu).$$

**Theorem 7.40** *If $f \in C(\bar{D} \times [0, +\infty))$ satisfies (7.25), (7.26), then there exists $\epsilon > 0$ such that (7.24) has positive solutions $u \in W^{2,p}(D) \cap W_0^{2,2}(D)$, for all $p \geq 1$, provided either*

(i) *$a > 0$ (possibly $+\infty$) in $D$ and $\lambda \in [\lambda_\infty - \epsilon, \lambda_\infty[$*

  *or*

(ii) *$A < 0$ (possibly $-\infty$) in $D$ and $\lambda \in ]\lambda_\infty, \lambda_\infty + \epsilon]$.*

In the remainder of this section, following Ambrosetti, Arcoya, and Buffoni (1994), we prove that

$$\lambda_\infty \text{ is a bifurcation from infinity.} \tag{7.28}$$

We remark that $u$ is a positive solution of (7.24) if and only if $u$ is a positive solution of

$$\Phi(\lambda, u) = 0,$$

where

$$\Phi(\lambda, u) := u - \lambda F(f(|u|)).$$

We rescale $\Phi$ as

$$\Psi(\lambda, z) := \begin{cases} ||z||^2 \Phi\left(\lambda, \frac{z}{||z||^2}\right) & z \neq 0 \\ 0 & z = 0. \end{cases}$$

Due to (7.26) the function $\Psi$ is continuous. It is clear that $\lambda_\infty$ is a bifurcation from infinity for (7.24) if and only if it is a bifurcation from the trivial solution $z = 0$ for $\Psi = 0$.

**Lemma 7.41** *For every $\lambda < \lambda_\infty$ there exists $r = r^-(\lambda) > 0$ such that*

$$\Phi(\mu, u) \neq 0 \ \text{for all } \mu \in [0, \lambda], ||u|| \geq r.$$

**Proof** Suppose, on the contrary, that there exist $\lambda^* < \lambda_\infty$, $\{\lambda_n : n \in \mathbb{N}\}$, $\{u_n : n \in \mathbb{N}\}$, such that

$$\lambda_n \to \lambda^*, \ ||u_n|| \to +\infty, \ u_n = \lambda_n F(f(|u_n|)).$$

Setting $w_n := u_n ||u_n||^{-1}$, elliptic theory and (7.26) yield that, up to a subsequence,

$$w_n \to w \ \text{in} \ C^{1,\alpha}(\bar{D})$$

for some $0 < \alpha < 1$, where $||w|| = 1$ and

$$\begin{cases} -\Delta w = \lambda^* m |w| & x \in D \\ w = 0 & x \in \partial D \end{cases}. \tag{7.29}$$

By virtue of the maximum principle, $w \geq 0$ and so $w = \gamma \varphi_1$, for some $\gamma > 0$. Since $||w|| = 1$, and in particular $w \neq 0$, we must have $\lambda^* m = \lambda_1$, which is in contradiction with the hypothesis $\lambda^* < \lambda_\infty$. $\qquad\square$

The proof of the following lemma can be found in Ambrosetti and Hess (1980).

**Lemma 7.42** *If $\lambda > \lambda_\infty$ there exists $r = r^+(\lambda) > 0$ such that*

$$\Phi(\lambda, u) \neq t\varphi_1 \ \text{for all } t \geq 0, ||u|| \geq r.$$

Setting $\lambda_n^- := \lambda_\infty - \frac{1}{n}$ and $\lambda_n^+ := \lambda_\infty + \frac{1}{n}$, we choose an increasing sequence $\{r_n\}$ such that $r_n \to +\infty$ and $r_n \geq \max\{r^-(\lambda_n^-), \ r^+(\lambda_n^+)\}$. If $\lambda \leq \lambda_n^-$, by Lemma 7.41 and Theorem 7.10

$$d\left(\Psi(\lambda, \cdot), B\left(0, \frac{1}{r_n}\right), 0\right) = d\left(\Psi(0, \cdot), B\left(0, \frac{1}{r_n}\right), 0\right)$$

$$= d\left(I, B\left(0, \frac{1}{r_n}\right), 0\right)$$

$$= 1. \tag{7.30}$$

On the other hand, by Lemma 7.42 and Theorem 7.10 and for all $t \in [0, 1]$,

$$d\left(\Psi(\lambda_n^+, \cdot), B\left(0, \frac{1}{r_n}\right), 0\right) = d\left(\Psi(\lambda_n^+, \cdot) - t\varphi_1, B\left(0, \frac{1}{r_n}\right), 0\right)$$

$$= d\left(\Psi(\lambda_n^+,\cdot) - \varphi_1, B\left(0,\frac{1}{r_n}\right),0\right). \quad (7.31)$$

We claim that

$$d\left(\Psi(\lambda_n^+,\cdot) - \varphi_1, B\left(0,\frac{1}{r_n}\right),0\right) = 0. \quad (7.32)$$

To this end, and by virtue of Theorem 7.8, it suffices to prove that the equation

$$\Psi(\lambda_n^+,\cdot) - \varphi_1 = 0$$

has no solutions on $\bar{B}\left(0,\frac{1}{n}\right)$. Indeed, suppose that if there is a function $z_n \in \bar{B}\left(0,\frac{1}{n}\right)$ such that

$$\Psi(\lambda_n^+, z_n) = \varphi_1.$$

Setting $u_n := ||z_n||^{-2}z_n$, we have $||u_n|| \geq r_n$ and

$$u_n - \lambda_n^+ F(f(|u_n|)) = ||u_n||^2 \varphi_1. \quad (7.33)$$

Due to (7.26), we may assume that

$$|f(|u|)| \leq C(1 + |u|) \text{ for all } u \in \mathbb{R}.$$

Equation (7.33) yields

$$-\Delta u_n - \lambda_n^+ f(|u_n|) = \lambda_1 ||u_n||^2 \varphi_1,$$

and, multiplying the latter equation by $\varphi_1$ and integrating by parts, we obtain

$$\lambda_1 ||u_n||^2 = \lambda_1 \int_D u_n \cdot \varphi_1 - \lambda_n^+ \int_D f(|u_n|) \cdot \varphi_1,$$

where we have used the fact that $||\varphi_1||_{L^2(D)} = 1$. Therefore, using Hölder's Inequality, we deduce that

$$\lambda_1 ||u_n||^2 \leq \lambda_1 (\text{meas}(D))^{\frac{1}{2}} ||u_n|| + (\lambda_\infty + 1)C(\text{meas}(D))^{\frac{1}{2}}(1 + ||u_n||). \quad (7.34)$$

Since $||u_n|| \geq r_n$ and $r_n \to +\infty$, it is easy to see that for $n$ large (7.34) cannot hold, and we deduce (7.32).

As the degree function takes only integer values, by (7.30), (7.31), and (7.32) we conclude that $\lambda \mapsto d\left(\Phi(\lambda,\cdot), B\left(0,\frac{1}{r_n}\right),0\right)$ is not continuous on the interval $[\lambda_n^-, \lambda_n^+]$, and so, by Theorem 7.10, there must exist $\lambda_n \in (\lambda_n^-, \lambda_n^+), z_n \in C(\bar{D}), ||z_n|| = \frac{1}{r_n}$, such that

$$\Psi(\lambda_n, z_n) = 0.$$

Thus, $u_n := \frac{z_n}{||z_n||^2}$ satisfies (7.24) for $\lambda_n$, $\lambda_n \to \lambda_\infty$, and $||u_n|| = r_n \to +\infty$. This concludes the proof of (7.28).

## 7.7   Exercises

**Exercise 7.1** Let $(X, ||\cdot||)$ be a normed linear space such that $\dim(X) = +\infty$. Prove that there exists a sequence $\{x_n\} \subset \partial B(0,1)$ such that $||x_n - x_m|| \geq 1$ for $n \neq m$.

*Solution 7.1.* Choose $x_1 \in X$ such that $||x_1|| = 1$ and let

$$X_1 = \text{span}\{x_1\} = \{\lambda x_1 : \lambda \in \mathbb{R}\}.$$

Since $\dim(X) > 1$, there exists $y \in X \setminus X_1$ and choose $z \in X_1$ such that $\rho(y, X_1) = |z - y| > 0$ and set $x_2 = \frac{y-z}{||y-z||}$. Then $||x_2|| = 1$ and

$$||x_2 - x_1|| = \left|\left| \frac{y - z}{||y - z||} - x_1 \right|\right| = \frac{1}{||y - z||}||y - z - x_1||y - z||\,||$$

$$\geq \frac{\rho(y, X_1)}{||y - z||} = 1.$$

Recursively, we construct $\{x_1, \ldots, x_l\} \subset \partial B(0,1)$ such that $||x_n - x_m|| \geq 1$ for $n \neq m$, $n, m = 1, \ldots, l$, and we set

$$X_l := \text{span}\{x_1, \ldots, x_l\} = \{\lambda_1 x_1 + \ldots + \lambda_l x_l : \lambda_1, \ldots, \lambda_l, \in \mathbb{R}\}.$$

Now repeat the same argument with $X_l$ instead of $X_1$, which is possible since $\dim(X) = +\infty$. We obtain a sequence $\{x_n\} \subset \partial B(0,1)$ such that $||x_n - x_m|| \geq 1$ for $n \neq m$.

**Exercise 7.2** Prove that there exists a constant $L > 0$ such that the following estimates hold: for every $u, v \in \bar{B}$ such that $u \neq v$,

$$|G_1(u,v)|_2 \leq \frac{L}{|u - v|_2^{N-2}}, \quad |G_2(u,v)|_2 \leq \frac{L}{|u, v|^{N-2}},$$

$$|\nabla_u G_1(u,v)|_2 \leq \frac{L}{|u - v|_2^{N-1}}, \quad |\nabla_u G_2(u,v)|_2 \leq \frac{L}{|u, v|^{N-1}},$$

and

$$|\nabla_u^2 G_1(u,v)|_2 \leq \frac{L}{|u - v|_2^N}, \quad |\nabla_u^2 G_2(u,v)|_2 \leq \frac{L}{|u, v|^N},$$

where $\nabla_u G_i$ and $\nabla_u^2 G_i$ are defined in the proof of Lemma 7.38.

*Solution 7.2.* Let $w_N = \mathcal{L}^N(B)$ be the measure of the unit ball in $\mathbb{R}^N$. By simple computation, we obtain that

$$\frac{\partial G_1(u,v)}{\partial u_j} = \frac{(u_j - v_j)}{Nw_N|u - v|_2^{N-2}}, \quad \frac{\partial G_2(u,v)}{\partial u_j} = \frac{(u_j|v|_2^2 - v_j)}{Nw_N|u - v|_2^{N-2}},$$

$$\frac{\partial^2 G_1(u,v)}{\partial u_j \partial u_k} = \left[\delta_{kj} - \frac{N(u_j - v_j)(u_k - v_k)}{|u - v|_2^2}\right] \frac{1}{N w_N |u - v|_2^N}, \qquad (7.35)$$

and

$$\frac{\partial^2 G_2(u,v)}{\partial u_j \partial u_k} = \left[\delta_{kj}|v|_2^2 - \frac{N(u_j|v|_2^2 - v_j)(u_k|v|_2^2 - v_k)}{|u,v|^2}\right] \frac{1}{N w_N |u - v|_2^N}, \qquad (7.36)$$

where $\delta_{kj}$ is the Kronecker symbol. Therefore, there exists a constant $L > 0$ such that for every $u, v \in \bar{B}$, $u \neq v$, we have

$$|G_1(u,v)|_2 \leq \frac{L}{|u - v|_2^{N-2}}, \quad |G_2(u,v)|_2 \leq \frac{L}{|u,v|^{N-2}},$$

$$|\nabla_u G_1(u,v)|_2 \leq \frac{L}{|u - v|_2^{N-1}}, \quad |\nabla_u G_2(u,v)|_2 \leq \frac{L}{|u,v|^{N-1}},$$

and

$$|\nabla_u^2 G_1(u,v)|_2 \leq \frac{L}{|u - v|_2^{N}}, \quad |\nabla_u^2 G_2(u,v)|_2 \leq \frac{L}{|u,v|^{N}}.$$

**Exercise 7.3** Prove that for every $u, v \in \bar{B}$ such that $u \neq v$, we have

$$\Delta_u G_i(u,v) = 0, \quad i = 1, 2$$

where $\Delta_u G_i := \displaystyle\sum_{j=1}^{N} \frac{\partial^2 G_i}{\partial u_j^2}$.

*Solution 7.3.* This follows immediately from (7.35) and (7.36).

# REFERENCES

Acerbi, E. and Fusco, N. (1984). Semicontinuity problems in the calculus of variations. *Archive for Rational Mechanics and Analysis*, **62**, 371–387.

Adams, R. A. (1975). *Sobolev Spaces*. Academic Press, New York .

Alexandroff, P. and Hopf, H. (1935). *Topologie*. Springer, Berlin.

Amann, H. (1976). Fixed point equations and nonlinear eigenvalue problems in ordered Banach space. *SIAM Review*, **18**, 620–709.

Ambrosetti, A., Arcoya, D., and Buffoni, B. (1994). Positive solutions for some semi-positone problems via bifurcation theory. To appear.

Ambrosetti, A. and Hess, P. (1980). Positive solutions of asymptotically linear elliptic eigenvalue problems. *Journal of Mathematical Analysis and Applications*, **73**, 411–422.

Ball, J. M. (1978). Convexity conditions and existence theorems in nonlinear elasticity. *Archive for Rational Mechanics and Analysis*, **63**, 337–403.

Ball, J.M. (1981). Global invertibility of Sobolev functions and the interpenetration of the matter. *Proceedings of the Royal Society of Edinburgh*, **88A**, 315–328.

Ball, J.M. and Murat, F. (1984). $W^{1,p}$ quasiconvexity and variational problems for multiple integrals. *Journal of Functional Analysis*, **58**, 225–253.

Calderón, A. P. (1951). On the differentiability of absolutely continuous functions. *Rivista di Matematica, Universita di Parma*, **2**, 203–213.

Carathéodory, C. (1914). Über das lineare Mass von Punktmengen eine Verallgemeinerung des Längenbegriffs. *Nachrichten Gesellschaft der Wissenschaften zu Goettingen*, pp.404–426.

Castro, A., Garner, J. B., and Shivaji, R. (1993). Existence results for classes of sublinear semipositone problems. *Results in Mathematics*, **23**, 214–220.

Ciarlet, P.G. and Nečas, J. (1987). Injectivity and self contact in non linear elasticity. *Archive for Rational Mechanics and Analysis*, **97**, 171–188.

Dacorogna, B. (1987). *Direct Methods in the Calculus of Variations*. Vol. 78. Springer-Verlag, Berlin; New York.

Dacorogna, B. and Fonseca, I. (1992). A minimization problem involving variation of the domain. *Communications in Pure and Applied Mathematics*, **XLV**, 871–897.

Davini, C. (1986). A proposal for a continuum theory of defective crystals. *Archive for Rational Mechanics and Analysis*, **96**, 295–317.

Davini, C. and Parry, G. (1989). On the defect-preserving deformations in crystals. *International Journal of Plasticity*, **5**, 337–369.

Deimling, K. (1985). *Nonlinear Functional Analysis*. Springer-Verlag, Berlin; New York.

de Rham (1955). *Varietes Differentiables*. Herman et Cie, Paris.

Dold, A. (1972). *Lectures on Algebraic Topology.* Springer-Verlag, Berlin; New York.

Dugundji, J. (1951). An extension of Tietze's theorem. *Pacific Journal of Mathematics,* **1**, 353–367.

Eisenberg, M. (1974). *Topology.* Holt–Rinehart–Winston.

Ericksen, J.L. (1987). Twinning of crystals I, pp. 77–96. In *Metastability and Incompletely Posed Problems.* S. Antman *et al.,* eds., IMA Vol in Applied Mathematics, No. 3.

Evans, L. C. and Gariepy, R. F. (1992). *Measure Theory and Fine Properties of Sobolev Functions.* Studies in Advanced Mathematics, Boca Raton.

Federer, H. (1969). *Geometric Measure Theory.* Springer-Verlag, Berlin; Heidelberg; New York.

Fonseca, I. and Gangbo, W. (1995). Local invertibility of Sobolev functions. *SIAM Journal of Mathematics,* **26** (2), 280-304.

Fonseca, I. and Müller, S. (1992). Quasiconvex integrands and lower semicontinuity in $L^1$. *SIAM Journal of Mathematical Analysis,* **23**, 1081–1098.

Fonseca, I. and Parry, G. (1987). Equilibrium configurations of defective crystals. *Archive for Rational Mechanics and Analysis,* **97**, 189–223.

Gold'sthein, V. M. and Reshetnyak, Y. G. (1990). *Quasiconformal Mappings and Sobolev Spaces.* Kluwer Academic Publishers, Dordrecht; Boston.

Gold'sthein, V. M. and Vodopyanov, S. (1977). Quasiconformal mappings and spaces of functions with generalized first derivatives. *Siberian Mathematical Journal,* **17 (3)**, 515–531.

Guillemin, V. and Pollack, A. (1974). *Differential Topology.* Prentice-Hall, Englewood Cliffs.

Hausdorff, F. (1919). Dimension und äusseres Mass. *Mathematische Annalen,* **79**, 157–179.

Havin, V. and Maz'ya, V. (1972). Non linear potential theory. *Russian Mathematic Surveys,* **27**, 71–148.

Hayman, W. K. and Kennedy, P. B. (1976). *Subharmonic functions.* Vol. I. Academic Press, London; New York.

Heinonen, K. and Koskela, P. (1993). Sobolev mappings with integrable dilatations. *Archive for Rational Mechanics and Analysis,* **125**, 81–97.

Heinz, H. (1959). An elementary analytic theory of the degree of mappings in $n$-dimensinal spaces. *Journal of Mathematics and Mechanics,* **8**, 231–247.

Hess, P. and Kato, T. (1980). On some linear and nonlinear eigenvalue problems with an indefinite weight function. *Communications in Pure and Applied Mathematics,* **5**, 999–1030.

Hopf, H. (1926). Abbildungsklassen $n$-dimensinaler mannigfaltigkeiten. *Mathematische Annalen,* **96**, 209–224.

Kuratowsky, K. (1966). *Topology I.* Academic Press, London; New York.

Lebesgue, H. (1907). Sur le problème de Dirichlet. *Rendiconti Circolo Matematico di Palermo,* **27**, 371–402.

Lloyd, N. G. (1978). *Degree Theory.* Cambridge University Press.

Malý, J. (1993). Weak lower semicontinuity of polyconvex integrals. *Proceedings of the Royal Society of Edinburgh*, **123A**, 681–691.

Malý, J. (1994). The area formula for $W^{1,N}$ - mappings. *Comment. Math. Univ. Carolinae*, **35**, (2), 291–298.

Manfredi, J. (1994). Weakly monotone functions. *Journal of Geometric Analysis*, **4**, (2), 393–402.

Manfredi, J. and Villamor, E. Mappings with integrable dilatation in higher dimensions. To appear in the *Bulletin of the American Mathematical Society*.

Marcus, M. and Mizel, V. (1973). Transformations by functions in Sobolev spaces and lower semicontinuity for parametric variational problems. *Bulletin of the American Mathematic Society*, **79**, 790–795.

Martio, O. and Ziemer, W. P. (1992). Lusin's condition $(N)$ and mappings with non-negative jacobian. *Michigan Mathematical Journal*, **39**, 495–508.

Meyers, N. (1970). A theory of capacities for potentials of functions in Lebesgue spaces. *Mathematica Scandinavica*, **26**, 255–292.

Morrey, C.B. (1966). *Multiple Integrals in the Calculus of Variations*. Springer-Verlag, Berlin; Heidelberg; New York.

Müller, S. (1990a). Higher integrability of determinants and weak convergence in $L^1$. *Journal fuer die Reine und Angewandte Mathematik*, **412**, 20–34.

Müller, S. (1990b). A remark on the distributional determinant. *Comptes Rendus de l'Académie des Sciences de Paris*, **311**, 13–17.

Müller, S., Tang, Q., and Yan, B.S. (1994) . On a new class of elastic deformations not allowing for cavitation. *Analyse Nonlineaire*, **11**, 217–243.

Nagumo, M. (1951). A theory of degree of mapping based on infinitesimal analysis. *American Journal of Mathematics*, **73**, 485–496.

Nirenberg, L. (1974). *Topics in Nonlinear Functional Analysis*. New York University.

Rabinowitz, P. H. (1971). Some global results for nonlinear eigenvalue problems. *Journal of Functional Analysis*, **7**, 487–513.

Rabinowitz, P. H. (1975). *Théorie du degré topologique et applications à des problèmes aux limites non linéaires*. Notes Univ. Paris VI et CNRS (written by H. Berestycki).

Rado, T. and Reichelderfer, P. V. (1955). *Continuous transformations in analysis with an introduction to algebraic topology*. Die Grundlehren der Math. Wissenschaften, Vol. 75, Springer-Verlag, Berlin.

Reshetnyak, J. G. (1969). On the concept of capacity in the theory of functions with generalized derivatives. *Sib. Mat. Zh.*, **10** (1969), 1109–1138 (Russian). English translation, (1969). Siberian Mathematical Journal, **10**, 818–842.

Reshetnyak, J. G. (1989). Spaces Mappings with Bounded Distortion. *Translations of Mathematical Monographs*, American Mathematical Society.

Rudin, W. (1966). *Real and Complex Analysis*, McGraw-Hill Book Company, New York.

Schwartz, J. T. (1969). *Nonlinear Functional Analysis*, Gordon and Breach, New York.

Smoller, J. and Wasserman, A. (1987). Existence of positive solutions for semilinear elliptic equations in general domains. *Archive for Rational Mechanics and Analysis*, **98**, 229–249.

Spivak, M. (1979). *A Comprehensive Introduction to Differential Geometry*, Vol. I. Publish or Perish Inc., Berkeley.

Stein, E. M. (1970). *Singular Integrals and Differentiability Properties of Functions*, Princeton University Press.

Šverák, V. (1988). Regularity properties of deformations with finite energy. *Archive for Rational Mechanics and Analysis*, **100**, 105–127.

Tang, Q. (1988). Almost-everywhere injectivity in nonlinear elasticity. *Proceedings of the Royal Society of Edinburgh*, **109**, 79–95.

Tartar, L. (1979). Compensated compactness and applications to partial differential equations. *Nonlinear Analysis and Mechanics*, Heriot-Watt Symposium, **IV**, R. Knops, ed., Res., Notes in Mathematics No. 39, pp. 136–212, Pitman, San Francisco.

Tartar, L. (1988). A short introduction to topological degree. Manuscript. Unpublished.

Varga, S. (1962). *Matrix Iterative Analysis*, (3rd edn). Prentice-Hall, Englewood Cliffs.

Ziemer, W. P. (1989). *Weakly Differentiable Functions*. Springer-Verlag, New York.

# INDEX

Absolutely continuous measure, 77
Admissible, 23
Algebra, 75
Antipodal point, 63
Approximate differential, 108
Approximately Hölder continuous, 101

Baire's Theorem, 69
Besicovitch's Covering Theorem, 98
Bessel capacity, 105
bifurcation from infinity, 200
Borel
  $\sigma$ algebra, 75
  measure, 75
  regular measure, 75
Borsuk–Ulam Theorem, 63
bounded distortion, 151
Brouwer
  degree, 17
  Fixed Point Theorem, 51

Capacity
  Bessel, 105
  linear, 105
  of a set, 92
Cauchy's Lemma, 42
Cauchy–Riemann equations, 44
Change of variables
  degree, 135, 138, 141, 145
  multiplicity, 132, 141, 145
Cofactors of a matrix, 26
Compact
  perturbation of the identity, 174
  mapping, 174
  perturbation of the identity, 177
  transformations, 178
Convex
  set, 48
  uniformly, 184
Crease, 6
Critical point, 5
Critical value, 5

Degree, 6, 14, 17, 23
  Brouwer, 17
  change of variables, 138, 141, 145
  Leray–Schauder, 177
Derivative of a measure, 76
Differentiable measure, 76
Differential
  approximate, 108
  weak, 108
Dilatation, 151
  finite, 151
Discrete, 145, 151
Domain decomposition property, 32, 182
Domain invariance, 184
Dugundji's theorem, 186

Excision property, 32, 182
Extension Theorem, 91

Fine cover, 83
Finite $\sigma$, 76
Finite dilatation, 151
First Version of Hopf's Theorem, 39
Fixed Point Theorem, 190
  Borsuk, 62
  Brouwer, 51
  Schauder, 185
Floquet solution, 53

Gauge, 48
Green operator, 200

Ham-Sandwich Theorem, 63
Hausdorff
  dimension, 81
  measure, 78
Homeomorphic, 64
Homotopy
  $C^0$, 30
  $C^1$, 15
  compact transformations, 178
  invariance, 179

Hopf's Theorem, 39
Hyperplane, 63

Index of a $p$-point, 34
Inequality
    isodiametric, 80
    Sobolev–Niremberg–Gagliardo, 90
Invariance
    domain, 184
    homotopy, 179
    of Domain Theorem, 68
Isodiametric inequality, 80
Isolated $p$-point, 33

Jordan Separation Theorem, 64, 185

Lebesgue
    density, 101
    outer measure, 74
    point, 77
Lebesgue–Besicovitch Differentiation
            Theorem, 77
Leray example, 172
Leray–Schauder degree, 177
Linear capacity, 105
Lipschitz
    boundary, 88, 145
    domain, 88

Mapping of bounded distortion, 151
Measurable, 76
    set, 75
Measure, 75
    absolutely continuous, 77
    Borel, 75
    Borel regular, 75
    derivative, 76
    differentiable, 76
    Lebesgue outer, 74
    outer, 74
    Radon, 75
    regular, 75
    restriction, 75
    s-dimensional Hausdorff, 78
    singular, 77
Minkowski function, 48
Mollifier, 17
Monotone mappings, 119

Multiplication Theorem, 35
Multiplicity
    change of variables, 132, 141, 145
    function, 106

Odd
    mapping, 55
    Mapping Theorem, 60, 184
Ordinary differential equation, 191
Outer measure, 74

Peano's Theorem, 70, 73
Periodic, 53
positone
    problem, 199
    semi, 199
Property
    $N$-property, 108
    $N^{-1}$-property, 108
    domain decomposition, 32, 182
    excision, 32, 182
    invariance of domain, 68
Pull-back, 22

Quasi-light, 151
Quasicontinuous function, 97
Quasiconvex function, 161

Radon measure, 75
Regular measure, 75
Regular value, 6
Restriction of a measure, 75
Riesz Representation Theorem, 77

Sard
    Lemma, 10, 21, 22
    Lemma for $C^1$ functions, 9
    Theorem for Sobolev functions, 110
Schauder Fixed Point Theorem, 185
Second Version of Hopf's Theorem, 39
semi-positone, 199
Set of finite dimension, 174
Singular measure, 77
Sobolev space, 87
Sobolev–Niremberg–Gagliardo
            Inequality, 90
Symmetric, 55

Theorem
  Baire, 69
  Besicovitch's Covering, 98
  Borsuk Fixed Point, 62
  Borsuk–Ulam, 63
  Brouwer Fixed Point, 51
  Differentiation for Radon
      Measures, 77
  Dugundji, 186
  Extension, 91
  First Version of Hopf's, 39
  Fixed Point, 190
  Ham-Sandwich, 63
  Homeomorphism, 185
  Invariance of Domain, 68, 184
  Jordan Separation, 64, 185
  Lebesgue–Besicovitch
      Differentiation, 77
  Multiplication, 35, 184
  Odd Mapping, 60, 184
  Peano, 70, 73
  Perron–Frobenius, 52, 71
  Poincaré–Bohl, 47

  Riesz Representation, 77
  Sard for $C^1$ functions, 9
  Sard for Sobolev functions, 110
  Schauder Fixed Point, 185
  Second Version of Brouwer Fixed
      Point, 51
  Second Version of Hopf's, 39
  Sobolev–Niremberg–Gagliardo, 90
  Tietze Extension, 16
  Vitali's Covering, 82
Tietze Extension Theorem, 16
Trace of a function, 92

Uniformly convex, 184

Vitali's Covering Theorem, 82

Weak differential, 108
Weakly
  differentiable, 108
  monotone mappings, 119
Winding number, 41